D1432977

The Plight and Promise of Arid Land Agriculture

The Plight and Promise of Arid Land Agriculture

C. Wiley Hinman

and

Jack W. Hinman

Columbia University Press

New York

Columbia University Press
New York Oxford

Library of Congress Cataloging-in-Publication Data

Hinman, C. W.
The plight and promise of arid land agriculture / C. W. Hinman and J. W. Hinman.
p. cm.
Includes bibliographical references and index.
ISBN 0–231–06612–0
1. Arid regions agriculture. I. Hinman, J. W. II. Title.
S613.H67 1992630′.9154—dc20 91–43989

Casebound editions of Columbia University Press books are Smyth-sewn and printed on permanent and durable acid-free paper.

Printed in the United States of America

c 10 9 8 7 6 5 4 3 2 1

To the improvement of the world's arid and
semiarid lands and the lot of those
dependent upon them.

Contents

Acknowledgments

We have received much welcome help from a great many people, and we wish to express our appreciation especially to the following:

first to our wives for editorial help, typing, and sacrifice during the research and writing of this book;

to Edward E. Lugenbeel for encouraging us to write this book;

to Dr. Gerald A. Boyack and Mary L. Hinman for valuable help with our computers and word processing; and

to Dr. L. H. (Burt) Princen and Dr. Kennith E. Foster our thanks for reviewing the manuscript and offering many useful suggestions.

The Plight and Promise of Arid Land Agriculture

Introduction

The first comprehensive overview of the complex forces influencing the world's soils and cropland base was published in 1976 by Eric Eckholm. He detailed the extent of soil erosion, deforestation, desertification, salinization of irrigation systems, and other problems threatening the world's cropland. Desertification is the term applied to humanmade degradation of land causing it to lose fertility and the capacity to grow crops or livestock. Drought can stimulate rapid desertification, but most scientists agree that changes in climate are not the reason vast areas of once arable land go out of production each year. The main causes of desertification are humanmade: overgrazing, overcultivation, deforestation, and unwise irrigation. All of these causes are intensified by a rapidly expanding population.

Desertification is not a new phenomenon; it has been going on since the beginning of recorded history. Sumeria, the earliest recorded civilization, developed on the southern flood plain of the Tigris and Euphrates rivers before 4000 B.C. Irrigation was used to grow crops on the rich alluvial soils and the area supported a population of 17 to 25 million people. But over the next 2,000 years, waterlogging and salinity rendered the land increasingly unfit for agriculture. At first the farmers switched from wheat to more salt-tolerant crops such as barley, but finally no crop could be grown. Cities which had been the birthplace of writing and mathematics were abandoned to decay. This area served as the battleground for both the recent eight-year (1980–1988) war between Iraq and Iran and the Persian Gulf war between Iraq and the United States and its allies. Two thousand years ago the fertile fields of northern Africa were the granary of the Roman Empire. Today much of this region is desert and more than half of the grain needed by the present population of North Africa must be imported. Similar examples of desertification and loss of agricultural productivity have continued from these early times to the present, and now on an increasing scale.

Desertification is not a problem restricted to the Middle East and North Africa; it is a worldwide problem. There are five major desert areas in the world, and near each of these is arid land (ten to twenty-three inches of rainfall per year). Humankind, by unwise actions, is pulling the desert into these areas at an alarming rate. One-third of the earth's land surface is arid or semi-arid. According to the United Nations Environment Program (UNEP), 11 billion acres, or 35 percent of the earth's land surface, is under direct threat of desertification.

In past centuries when an area was destroyed for habitation, having lost its capability to provide food and shelter, the population could migrate to an uninhabited region or displace a weaker or less sophisticated population in a desirable location. Today, with the world's population exceeding 5 billion and expanding at the rate of some 90 million per year, migration is no longer a viable solution to the problem. From 1968 through 1973 the Sahel region of West Africa suffered a disastrous drought killing thousands of people and millions of animals. This tragedy was the impetus for the convening of the United Nations Conference on Desertification (UNCOD) in Nairobi, Kenya in 1977 to prepare a plan of action to combat desertification. The plan, as approved by representatives of ninety-five nations, consisted of twenty-eight recommendations to halt desertification. From 1980 to 1984 Africa was stricken with another terrible drought. Soils could no longer support agriculture, forests were destroyed, famine and political pressures caused massive relocation of people. However, desertification is a low-priority item for politicians. The plan suffered from inadequate funding and little real commitment by the governments involved.

In 1984 UNEP found that little progress had been made and the problem was rapidly worsening.

In addition to desertification, severe erosion, waterlogging, and salinization of irrigated land, there are nonagricultural uses that contribute significantly to loss of cropland. A growing population's land needs include urbanization, industrialization, mining, energy production, and transportation. The worldwide growth of cities is a major cause of cropland loss. In the United States between 1967 and 1975, urban sprawl consumed 6.2 million acres of prime cropland. Similar losses of cropland to urbanization are occurring globally and are not limited to large cities. For example, in India there are some 600,000 villages and most of them are expanding. Each year 14 million people are added and to a large extent the living space required for them is built on crop land.

Rapid population growth, especially in Third World countries, and massive urbanization in these regions have led to vastly increased energy consumption and the requirement of sophisticated transportation systems. Whether the energy need is supplied by fossil fuels or from hydroelectric dams, valuable cropland is invariably lost. Hydroelectric dams usually inundate cropland, and oil refineries and storage tanks are almost always located on farmland; strip mining of coal leads to the long-term loss of farm and/or grazing land. Transportation systems require millions of acres of land for roads, streets, parking spaces, filling stations and other service facilities, and airports with hangars and terminals. In most cases the land utilized for these transportation needs was once farmland. Factories, shopping centers, and mills are usually built on the periphery of cities on land previously used for crops. The amount of cropland that will be developed for building, paved over, flooded, or strip mined between now and the end of the century will undoubtedly proceed at a rate at least as great as during the last quarter century.

Many people, including most political leaders, regard current surpluses and the dramatic success of the Green Revolution, as adequate safeguards against global famine in the future. The Green Revolution, which introduced new varieties of wheat and rice to Asia and Latin America along with fertilizers, pesticides, and mechanized farm equipment, increased harvests markedly, and improved the lives and prospects of millions of people. However, this agricultural progress was not distributed evenly. Millions of subsistence farmers in the Third World, raising their food on marginal, rain-fed land, did not benefit from the new varieties and technologies. Few of these farmers can afford the fuel and machinery required, and the crops subsistence farmers grow have received comparatively little research attention. In the long term, if the cropland losses in the developed countries caused by salinization, erosion, deforestation, urbanization, and industrialization continue at the present rate, crop surpluses even in the developed nations will cease to be a problem.

The bad news does not end here. In July 1986 a team of scientists studying the increasing level of carbon dioxide and other gases in the atmosphere reported solid evidence that the predicted global warming, or so-called greenhouse effect, had begun. By 1989 information on, and the resulting controversy over the greenhouse effect was appearing almost daily in the news. In February 1990 the Union of Concerned Scientists (UCS) presented President Bush with an "Appeal by the American Scientists to Prevent Global Warming." This document, which was signed by 52 Nobel laureates and over 700 members of the National Academy of Sciences, expressed the scientific community's consensus that global warming had emerged as the most serious environmental threat of the twenty-first century, a threat to raise the sea level, bring droughts to some areas and floods to others, and create unprecedented agricultural devastation. Global warming, oil spills, acid rain, and air pollution all have the same basic cause: the use of fossil fuels—coal, oil, and gas—to produce energy.

Between 1950 and 1986 global consumption of fossil fuels increased fourfold, for which most of the increase in the carbon dioxide level in the atmosphere is responsible. In addition, there is a marked increase in the oxides of nitrogen and sulfur. These oxides combine with rain and snow to return to earth as toxic acid (acid rain), which has inflicted incalculable damage to forests, streams, and lakes. Since 1980 forest damage has spread rapidly through central and northern Europe, involving more than 46 million acres, including the famed Black Forest of West Germany. Extensive forest injury has been reported also for Czechoslovakia, Austria, Yugoslavia, Switzerland, Romania, and Sweden, where dying lakes first brought attention to the problem of acid rain. Forest destruction on the scale occurring in Europe is not yet in evidence in North America, but trees are suffering nonetheless from air pollution. Damage is most severe in the high elevations of New York, Vermont, and New Hampshire. In one study on Camel's Hump in the Green Mountains of Vermont, researchers found that seedling production, tree density, and basal area declined about one-half between 1965 and 1979. Acid rain has inflicted serious damage upon numerous lakes and streams in Scandinavia and eastern North America. And many more are vulnerable as long as acid rain persists. Acidification of soils is occurring rapidly in heavily polluted industrial regions, including extensive areas of Eastern Europe that now resemble wasteland.

Another problem results from atmospheric pollution by chlorofluorocarbon compounds (CFCs) widely used in refrigerators and air conditioners, as propellants in aerosol products, cleaning fluids for electronic equipment, and for many other uses. These compounds, in addition to the gases produced in the burning of fossil fuels, are thinning the earth's protective shield of ozone in the upper atmosphere, allowing more of the sun's ultraviolet radiation to reach the earth's surface, thus increasing the danger of skin cancer, impairment of the immune system, and retardation of crop growth. Thousands of other chemicals are emitted from factories, applied to the soil as pesticides, fertilizers, and herbicides, and deposited in landfills and waterways as waste. Researchers are just beginning to assess the long-term risks these substances and practices present.

The plight is serious and its long-term consequences are global. The future of all agriculture is threatened by these problems, but agriculture, and arid land agriculture in particular, offers the potential to solve many of them. The first two chapters detail some of these dangers as they exist in specific areas of the world, while the rest of the book is devoted to describing emerging agricultural technologies, potential new food crops, plants which can provide alternative fuels and chemicals, the methods and processes involved in utilizing such crops, and finally, how governments, academia and the private sector can work together to achieve the promise these opportunities offer to the arid regions and to the entire world.

SELECTED INFORMATION SOURCES

Brown, L. R. 1978. *The Worldwide Loss of Cropland.* Worldwatch Paper 24. Washington, D.C.: Worldwatch Institute.

Brown, L. R. et al. 1986. *State of the World—1986.* New York: W. W. Norton.

Brown, L. R. et al. 1987. *State of the World—1987*. New York: W. W. Norton.

Caufield, C. 1987. "Fowl Play—Earth" *Omni* 9 (June):22.

Eckholm, E. 1976. *Losing Ground: Environmental Stress and World Food Prospects*. New York: W. W. Norton.

Eckholm, E., and L. R. Brown. 1977. *Spreading Deserts—The Hand of Man*. Worldwatch Paper 13. Washington, D.C.: Worldwatch Institute.

Environmental Defense Fund. 1987. 20th Anniversary Report, p. 13.

Grainger, A. 1982. *Desertification—How People Make Deserts, How People Can Stop and Why They Don't*. Nottingham: Russel Press.

Karrar, G., and D. Stiles. 1984. "The Global Status and Trend of Desertification." *Journal of Arid Environments* 7:309–312.

Pimentel, D., and C. A. Edwards. 1982. "Pesticides and Ecosystems." *BioScience*, July-August.

Postel, S. 1984. *Air Pollution, Acid Rain and the Future of Forests*. Worldwatch Paper 58. Washington, D.C.: Worldwatch Institute.

Reiger, G. 1987. "It Can't Happen to Me." *Field and Stream* 92 (July):34–37.

Reiger, G. 1987. "Don't Blame Me." *Field and Stream* 92 (August):19.

Schindler, D. W., et al. 1985. "Long Term Ecosystem Stress: The Effect of Years of Experimental Acidification of a Small Lake." *Science*, June 21.

Vogelmann, H. W. 1982. "Catastrophe on Camel's Hump." *Natural History*, November.

ONE

The Plight of Agriculture in the Eastern Hemisphere

Africa

Africa is the second largest continent, with an area of nearly 12 million square miles. Two-thirds of the continent lie north of the equator with the southern extremity at about the same latitude as Buenos Aires. Its northernmost extremity has a latitude slightly south of that of Washington, D.C. Until recently it was common to consider Africa as a continent of low population density and abundant availability of land. While it is true that the area is large, much of it suffers from low, unreliable rainfall and poor soils. The human population tends to be clustered around water supplies, roads, and the areas with the most fertile soil. While the population density may be low in absolute terms, it is high in terms of the limited carrying capacity of the land. Large areas of land are not suitable for human habitation because of tsetse flies, river blindness, rock outcropping, and desert.

The northern part of Africa—Morocco, Algeria, Tunisia, Libya, and Egypt—is primarily a lowland of which the Sahara Desert is the largest region. The population of arid North Africa has multiplied sixfold since the turn of the century. Land degradation in

Morocco, Algeria, Tunisia, and Libya accelerated after 1930 when the population of the region began to grow rapidly. The agricultural land was stressed by overgrazing, the extension of grain farming into marginal lands, and excessive firewood gathering (the chief source of heat for cooking and warmth in Africa is wood).

Egypt is also suffering from rapid population growth. For thousands of years Egyptian farmers irrigated by simple diversions from the Nile River, and until recently Egypt was self-sufficient in grain production, but by 1986 over half of the grain consumed was imported. More intensive irrigation made possible by the Aswan High Dam upset the long-standing water balance in some areas leading to severe waterlogging and salinity. Urban encroachment on the narrow strip of arable land along the Nile caused a further loss of cropland. Even further loss has resulted from the use of topsoil for building bricks. While drought and famine on the southern side of the Sahara made news during the 1970s and 1980s, land degradation along the northern margins of the Sahara went unpublicized. Human suffering was minimized because these countries received significant proceeds from petroleum and phosphate exports. In addition, millions of North Africans migrated to Europe for work and remitted sizable portions of their earnings to their native lands. Nonetheless, the stark truth is that this one-time granary of the Roman Empire is now a major food-importing region.

Sub-Saharan Africa, twice the size of the United States, is made up of forty-five different countries, and contains a wide range of climates and environments as well as a diversity of cultural, economic, and political characteristics. The population of the continent as a whole grew from 139 million in 1940 to 400 million by 1984. About 70 percent of the population live in rural areas and are predominantly subsistence farmers and herders. These are the people who produce most of Africa's food and in much of the region women play a major role in food production.

In recent years, the term Sahel has been applied to the group of countries bordering the Sahara Desert on the south: Senegal, Mauritania, Mali, Niger, Chad, Burkina Faso (formerly Upper Volta), Cape Verde, Guinea Bissau, and the Gambia. These countries, among the poorest in the world, suffered severely as a result of the 1968–1973 drought. They have banded together in an organization known as CILSS (the French acronym for the Permanent

Interstate Committee for Drought Control in the Sahel) to improve socioeconomic conditions and to minimize the adverse effects of future droughts. Both Sahelians and the international relief organizations realized that a long-range effort was needed to avoid future crises. In addition, the United States, mainly to alleviate human suffering, but also to expand stable markets for U.S. products and to maintain the availability of critical and strategic materials, created the Sahel Development Program (SDP) in 1977 as an amendment to the Foreign Assistance Act. International relief efforts provided over $360 million emergency relief to the Sahel by 1974, while U.S contributions to the Sahel development effort have amounted to $1.4 billion between 1976 and 1986. The U.S. support has been in cooperation with Club du Sahel (an association of donors and Sahelians) and the CILSS.

Many other organizations, both government and private, have given aid to this stricken region. The United Nations alone has some thirty-five agencies and programs related in some way to Sahelian assistance, with funding ranging from a few thousand to tens of millions of dollars. France is the largest single donor, while the European Community (formerly the European Economic Community [EEC]) is the second largest source of aid for the Sahel. U.S. aid to the Sahel has increased steadily since the early 1970s, making it the third largest donor over the 1978–1980 period. The World Bank contributes over $100 million per year. Saudi Arabia, Kuwait, the African Development Bank, Canada, and Japan all made significant contributions. In addition to these large-scale programs sponsored by individual governments and multilateral bodies, there are a number of government-backed and private groups whose collective contributions of both money and volunteer services have increased the total amount of aid significantly. These organizations include the Peace Corps, Catholic Relief Services, the World Council of Churches, Oxfam, and Save the Children. Unfortunately, all of this effort has provided only modest tangible success.

Increased agricultural production and environmental stabilization have not been achieved. Nevertheless, Sahelians and donors claim they have learned important lessons which may lead to more successful efforts in the future. It is apparent now that agricultural technologies suitable for the Sahel must be low-cost, low-risk, capable of providing substantial production increases, and be sustainable. Sahelian farmers and herders, including women, need to be consulted and given a more active role in the planning and execution of these efforts. Misguided policies on the part of the Sahelian governments and many of the donors have been an important factor in discouraging increased food production and effective distribution.

During the last twenty years the decline in per capita food production in the Sahel has required increased imports of cereals. At the time these nations gained their independence, food imports were negligible, but by 1985 imports provided one-third of the total cereal consumed. Because of the abject poverty of the people involved the gap between food production and consumption was not met by commercial imports but filled by food aid. In 1985 half of the cereal imports were grants in aid. Most of the imported food, which was surplus production from other parts of the world, was rice and wheat, and these grains soon became the staple foods in the large cities of the Sahel. Millet and sorghum remain the staples in most rural areas, but rice is becoming popular because it is easier to prepare and store. This changing demand further reduced the incentive for farmers to grow sorghum and millet while their climate and soil prevent them from growing wheat and rice. If existing trends continue, one-third of the cereals consumed in the Sahel will continue to be imports or food aid even in years of normal rainfall.

Desertification is a serious concern in the Sahel, especially in the northern and middle zones where the population exceeds the sustainable level and where tree cover, grasslands, and soils have all been degraded. In the Sahel fuel wood provides 85 percent of the region's energy. Commercialization of fuel wood and charcoal has resulted in far more trees being consumed than are being regenerated in spite of a variety of reforestation efforts. In six of the Sahelian countries fuel wood need exceeds available supplies to meet even minimum requirements.

Grassland degradation is the result of a number of factors including excessive numbers of livestock, concentrating herds near deep wells, and restrictions on herders' mobility imposed by the spreading of farms into traditional grazing areas. Removal of vegetative cover through deforestation and degradation of grassland has caused serious loss of fertility and has made the soil vulnerable to erosion.

Misguided policies with regard to cereal pricing, artificial exchange rates, poor debt management, low investment in food crops, and corruption in food aid distribution have all contributed to the poor agricultural performance in the Sahel. If future efforts in the Sahel are to be successful, more appropriate technologies and better development methods are essential. The first objective of the CILSS was to attain self-sufficiency in food production in the Sahel, but after more than a decade there is little evidence to indicate that this objective has been achieved.

The food crisis and degradation of the environment are not limited to the Sahelian nations of Africa. Even though most parts of Africa received normal or near-normal rainfall in 1985, there were still some 19 million people dependent on famine relief from Western Europe and North America. The United Nations Food and Agriculture Organization (FAO) estimated that at the beginning of 1986 Africa still needed over 8 million tons of imported grain to feed its population during the next few months. Emergency food deliveries were required to prevent starvation in the Sudan, Ethiopia, and Mozambique. The drought and famine reached its nadir in late 1984. Some 30 to 40 million people suffered from the famine; thousands died and millions were displaced from their homes. In Ethiopia, Angola, Chad, Mozambique, and Sudan the problem was aggravated by civil unrest. By the beginning of 1986 the drought had ended in eight countries (Burundi, Kenya, Lesotho, Morocco, Rwanda, Tanzania, Zambia, and Zimbabwe) that had been on the FAO emergency list. At the same time the United Nations Office for Emergency Operations in Africa (OEOA) estimated that 19.2 million Africans still suffered from the famine or its effects upon health, water supply, or agricultural potential. The refugee problem reached critical proportions in Somalia, Sudan, and Ethiopia and civil strife was rampant.

In the autumn of 1986 Africa was threatened with the recurrence of a problem which had been quiescent since the 1960s: the locust and grasshopper plagues. Grasshoppers and four species of locust constituted a serious threat to Africa during the latter part of the colonial period. When most of the African territories were achieving independence, the colonial locust-control agencies were supplanted by a combination of regional and national pest-control agencies of the newly independent nations. UN bodies provided support and during the desert locust plague of the time, control operations were regarded as increasingly effective. During the generally dry years after the mid-1960s there were no serious locust outbreaks and this caused a withering of the control apparatus. Following the rains of 1985 there were better harvests in many areas, but also a bigger crop of locusts and grasshoppers, threatening a return to the plagues of the past.

In mid-September of 1986 grasshoppers threatened crops in the broad area of West Africa. Four American DC-7s sprayed malathion over the threatened areas in Senegal, Mali, Mauritania, and the Gambia, successfully averting serious crop losses. This program, using planes under contract to the U.S. Agency for International Development (AID), was the largest single operation to control outbreaks of locusts and grasshoppers in a score of sub-Saharan countries. During the locust invasion of Mauritania in the fall of 1988, the control teams were unable to enter the northern part of the country because of the left-wing Policario guerrillas, who shot down one of the DC-7s, killing all five American crew members. Ethiopia and Sudan were reported to have serious locust infestations which could become major threats, but because of the civil wars, access by anti-locust forces was prohibitively hazardous. However, the Desert Locust Control Organization serving East Africa is regarded as the regional organization best prepared to deal with the situation. While information on this particular problem is sketchy because of the civil strife, it is apparent that even when rainfall is adequate there can still be famine.

It should be clear from this brief review that the problems of Africa are substantial and they are likely to worsen in the immediate future. Food production in Africa on a per capita basis has declined steadily during the last twenty years while population growth has risen to the highest rate in the world. Decision makers apparently have not been convinced of the seriousness of this problem when confronted with arguments of ecological and environmental deterioration. Perhaps, when costs in economic terms are addressed and considered in the light of potential long-term benefit to be gained from projects devoted to environmental conservation, governments and donor agencies will begin to give higher priority to the conservation and rehabilitation measures that are required.

Australia

The island continent of Australia is in the southern hemisphere southeast of Asia and between the Indian and Pacific oceans. Australia's land mass is about 2,400 miles east and west, nearly 2,000 miles north and south, an area almost as large as that of the United States. There are mountains and plateaus with the Great Dividing Range (its highest peak is Mt. Kosciusko, with an elevation of 7,316 feet) roughly paralleling the east and southeast coast and forming the watershed between streams flowing to the Coral Sea and the Pacific Ocean and those flowing to the Indian Ocean. There are lesser mountain ranges in the west, north, and center, but most of the continent is a plain. Most of the central and western areas are arid or semiarid with few permanent streams. The eastern, southeastern, and southwestern sections of Australia receive the most rainfall and are the most productive in terms of agriculture. These regions have some permanent streams, the best seaports, and are rich in forests and minerals. Almost 40 percent of the 1,621,200 square miles of Australia is desert made up of Acacia shrublands, hummock grasslands, tussock grasslands, and shrub steppe. Adjoining the desert are some 425,000 square miles of semiarid land made up of low woodlands, shrub woodlands, and eucalypti shrub woodlands. These are low mountain ranges and associated plateaus covering 224,000 square miles with elevations up to 5,900 feet. The mountains are never snow-covered, and rainfall runoff is immediate and short-lived. Rainfall is extremely variable both seasonally and annually. Along the eastern coastline most of the small rivers are ephemeral and the larger rivers suffer marked fluctuations in flow. With the exception of the Darling-Murray system, which draws its water from the eastern divide, there are no permanent rivers in the interior. There are large reserves of groundwater in the sedimentary basins of arid Australia, and these are the major sources of water for livestock and domestic use. The quality of the water varies considerably. The presence of calcium, sodium, sulfate, carbonate, and fluoride often imposes restrictions on the use of the groundwater. When used on cropland, salt accumulation is often a problem. For the most part, the lakes in the desert are saline or nonsaline playas, i.e., basins that temporarily become shallow lakes after heavy rains. Summer temperatures are high in the arid areas and surrounding territory with a mean of 86 degrees F. with little variability between and within the regions. Winter temperatures are less uniform with the mean temperature ranging from 52 degrees F. in the south to 66 degrees F. in the north. Winds tend to be light with low erosion potential over most of the desert except in the southwest Simpson Desert where strong winds blow in the opposite direction of the prevailing winds causing some mobile dune fields.

The flora of Australia is represented by large numbers of species within a few genera. For example, species of eucalyptus and acacia are numerous in both the desert areas and in the humid southeast portion of the continent. As a result of the extreme variability in precipitation, the perennial plants of Australia are long-lived, with irregular flowering, fruiting, and germination. With European colonization in the nineteenth century, Mediterranean-type annual plants were introduced, especially around the southern semiarid margin. Cattle and sheep were introduced and the famous rabbit invasion occurred in the 1890s. Today there is a small resident rabbit population in the desert south of the Tropic of Capricorn, but the major use of the plant resources is raising sheep in the south, southeast, and southwest, and cattle in the central and northern regions. Other uses of the arid lands include mining, aboriginal homelands, tourism, national parks, and military and telecommunication facilities.

Australia's human population is about 17 million, or 6 percent of the present U.S. population. Approximately 80 percent of the people live within 125 miles of the coastline and more than 85 percent are urban dwellers. The percentage of rural residents fell from 21.3 percent in 1954 to 14.4 percent in 1971 and that trend is continuing. In marked contrast to Africa, Australia has almost attained zero population growth and produces agricultural and pastoral products in excess of domestic needs. Currently there is a move away from cereal growing in favor of Merino sheep production.

Compared with many other arid regions of the world, the largely roadless, waterless, unoccupied deserts of Australia represent a stable habitat. The semiarid regions have suffered damage attributable to the introduction of herbivores and overgrazing, but the worst of this is in the past. However, current research indicates that overgrazing during droughts has caused considerable degradation of arid and semiarid lands. For example, in Queensland regions with subsoils of sodic clays and high salt levels, removal of vegetation by overgrazing, fire, and drought has led to a problem called scalding. When the topsoil is eroded to expose sodic clay subsoil, an impermeable crust of fine silt cemented together with sodic clay restricts entry of water into the subsoil. Seepage salting is another form of induced salinization. This results in cleared areas where water previously used by trees raises groundwater levels close to the surface. Water rises to the surface by capillarity and evaporation concentrates the salts at the surface. Fortunately, Australian agriculture professionals seem to have a good understanding of these problems, so catastrophic degradation of grazing and cropland on a large scale seems unlikely for now.

Eurasia

Europe and Asia constitute the largest land mass on earth with the smallest percentage of arid and semiarid land. However, this does not mean there are no desertification problems in this part of the world. In western Europe there is little true desert, but portions of Spain, Italy, and Greece are arid or semiarid; however, these are not the areas of greatest concern. The most important causes of farmland and forest destruction in central and northern Europe are acid rain and air pollution. As noted above, 1986 estimates place the area damaged at over 47 million acres or 14 percent of Europe's forests, an area the size of Austria and eastern Germany combined. Scientists in the affected countries continue to search for the cause of this unprecedented forest damage, but most agree that air pollutants such as sulfur dioxide, nitrogen compounds, ozone, and heavy metals in combination with natural factors, such as drought, insects, or cold, are the most likely cause.

Adverse changes in soil chemistry present another serious concern for European countries. For example, evidence of soil alteration has been reported in Eastern Europe. In the Erzgebirge Mountains northwest of Prague, Czechoslovakia forests have been severely damaged. Power plants in the region burn high-sulfur coal and sulfur dioxide concentrations average 112 micrograms per cubic meter, a level thirteen times higher than the concentrations found in undamaged rural forests to the southeast. Detailed measurements made on the runoff from the Erzgebirge Mountain watershed suggest that acidification of the soil has altered its ability to support a forest. The economic loss resulting from forest damage is substantial. The Czechoslovakian Academy of Science estimated the cost of acid pollution to be $1.5 billion per year. In addition to supplying lumber, forests provide protection from erosion of soil, protection for the quality of streams and groundwater, and recreational enjoyment for both residents and tourists. Researchers at the University of Berlin estimate the total cost of forest damage in the former West Germany to average $2.4 billion per year.

Poland has extensive areas of environmental disruption caused by acid rain, toxic pollution, and depletion of oxygen from its rivers. In upper Silesia uncontrolled sulfur dioxide and particulate emissions threaten both the land and the populace. The forests around Katowice suffer from the effects of ozone, sulfur dioxide, and hydrocarbon emissions. In southwestern Poland one-third of the forest area has been damaged.

In April 1986 the tragic Chernobyl nuclear power accident presented a new threat to the land and people by depositing health-threatening levels of radioactive materials more than 1,200 miles from the plant and in at least twenty different countries. In the area immediately surrounding Chernobyl the Soviets mounted a massive cleanup effort. All nearby forests were destroyed and the topsoil removed and buried. The farmland in the area was abandoned and probably will remain so for many decades. In the weeks following the accident fresh vegetables, berries, dairy products, beef, mutton, and freshwater fish in many parts of Europe were found to contain levels of radioactivity in excess of limits recommended by health authorities. The European Community (EC) adopted interim limits on radiation levels in crops and restrictions on the importation of fresh food from Eastern Europe for three weeks. Scientists believed that the radiation levels would fall sufficiently in most areas to permit a return to normal life within a year or two, although some problems would persist. For example, authorities banned fishing on Lake Lugano in Switzerland when they learned that fish from the lake contained levels of radioactive cesium up to six times the EC standard. Since cesium 137 has a half-life of thirty years, radioactive materials will probably remain in the lake sediments and some types of agricultural soils for many years. In Scandinavia the Lapps were severely affected. Nearly all of the reindeer slaughtered in late summer of 1986 showed radioactivity above the level considered safe for human consumption. Biologists fear that this situation may persist for years, imposing a major catastrophe for the fragile Lapland ecosystem and the people who depend on it.

The agricultural plight of the USSR is not limited to the problems caused by Chernobyl. In 1970 Soviet agriculture was able to meet all domestic needs and, in addition, export 8 million tons of grain. By 1981 the Soviet Union had to import 43 million tons of grain to make up for its domestic food deficit. The ability of a country to produce food is determined mainly by its natural resources and by how wisely it manages those resources. Among the world food producers, the Soviet Union is in a class by itself in terms of cropland area—currently over 500 million acres (as compared to 350 million acres planted annually in the United States). Differences in temperature and rainfall offset to some extent the advantages the Soviets have in area of arable land, because much of their farmland lies farther north (mostly between 48 and 55 degrees compared with 34 and 45 degrees north latitude for the US). In addition, the Soviet cropland which has a longer growing season in the south central part of the nation is largely semiarid. This has been a source of frustration for the Soviet agricultural planners and has led to an intense effort to develop irrigation capability in the semiarid regions.

Soviet irrigation systems have been plagued with many problems, including waterlogging and salinization. Some of these problems are the result of faulty design in construction. In some cases construction teams pressed to meet Soviet-planned goals have omitted drainage systems, and this can lead quickly to waterlogging and salinization. In Soviet Central Asia during the early 1960s, irrigated land abandoned due to these problems equaled the acreage of new land brought under irrigation. The situation improved markedly during 1971–1975 with newly irrigated land exceeding abandoned irrigated land by four to one. For many years Soviet planners considered schemes to divert the Ob River from its northerly course into the Arctic Sea and redirect it some 1,500 miles into the steppes of Central Asia. Another plan would divert the Sukhona River into the Volga, but these ambitious plans apparently have been shelved until at least the twenty-first century. As a consequence, the Aral Sea will continue to drop at the rate of 11.5 feet per year because of irrigation withdrawals from the Syr Dar'ya and Amu Dar'ya Rivers.

In 1973 the Aral Sea was the fourth largest inland body of water in the world, but its level has dropped nearly 40 feet since 1960. The Aral port of Muynak was at one time a great fishing center with more than 10,000 fishermen who provided about 11 percent of the Soviet catch. Now Muynak is 30 miles from the water line and there is no commercial fishing because salt concentrations in what is left of the lake have nearly wiped out the fish population. The once-thriving port now lies on the edge of a saline desert. Wind storms sweep up salt and sand and deposit the toxic mixture on the surrounding irrigated farmland. In addition to the salt damage to the crops, the moderating influence of the Aral Sea on the weather has decreased, making summers hotter and winters colder. The resulting crop declines are of national concern to the Soviet Union because this region used to provide over 90 percent of the nation's cotton, 30 percent of its rice and 25 percent of the fruits and vegetables. There are human health problems including unusually high rates of stomach and liver disease, throat cancer, and birth defects.

According to western observers, the long-term deterioration of agricultural production in the Soviet Union cannot be blamed on adverse weather conditions alone, as Soviet officials are inclined to do. Since 1966 the Soviet investment in agriculture has dwarfed that of the United States. In spite of this, Soviet agriculture has deteriorated on many fronts. Grain production peaked in 1978 at 229 million tons. By 1982 the total Soviet cereal harvest was down to about 170 million tons, and from 1987 to 1989 the harvest ranged from 186 to 203 million tons. Since 1978 there has also been a decline in the production of meat, milk, vegetables, and other agricultural products. The Soviet grain area has decreased in eight of the last nine years, resulting in a 12 percent decline overall. Some of this decrease is explained by an increase in fallowed land, but most of that represents abandonment of cropland. The 1986 harvest of durum wheat was purported to be a record for the five-year period 1981–1985, but was still 40–50 million tons short of what the country needed.

One of the underlying problems of Soviet crop production is soil erosion. Among those working the state and collective farms there is a lack of a strong sense of stewardship, which contributes to a severe erosion problem. This is aggravated by the use of large heavy equipment on vast fields, stimulating both wind and water erosion. Soviet soil scientists in the early 1980s pleaded with the agricultural bureaucracy to stem the loss of topsoil. Each year between 1.2 and 3.7 million acres of cropland were abandoned because it was so severely eroded that it was no longer worth farming. It is interesting to note that in 1981 Mikhail S. Gorbachev, then a Politburo member, urged the planners to heed the advice of the soil scientists, but in the short-term, efforts to boost production, responsible soil management practices were ignored. Thus, as of the mid-1980s, the Soviet Union led the world in net imports of meat as well as grain.

Water and air pollution also are problems of serious concern to the Soviet Union. It is reported that even in the early 1970s it was difficult to find an unpolluted river in the USSR. Fish kills had occurred all over the country including the Volga, Ural, and Dnieper rivers. Sulfur dioxide and nitrogen oxide emissions approached lethal levels in some areas and toxic emissions of lead and fluoride were permitted in large quantities. In one city in eastern Karakhstan lead concentrations reached fourteen times the maximum permissible level. Forests in the Ukraine have nearly disappeared. Disruptive strip mining of coal has proliferated and natural areas including Lake Baikal have been exploited and seriously disrupted. During the 1960s some fifty factories were built along the shores of Lake Baikal and most of them discharge their raw wastes into the water. An overview of the agricultural plight in the USSR shows, as Marshall I. Goldman concluded, "that public greed and lust can be as destructive as private greed and lust."

The only eastern bloc country to achieve food self-sufficiency is Hungary. In the 1960s Hungary was faced with mounting debts to the West, a grain production deficit requiring importation of several hundred thousand tons of grain per year and a deteriorating centralized agricultural system. In 1968 Hungry took the first steps to trim waste and inefficiency. They based domestic prices on world prices and began to decentralize agriculture by converting cooperatives formed in the original collectivization into self-managing cooperatives. This gave farmers incentive to work hard and efficiently and to produce crops that were needed. This shift to a market-oriented system revitalized Hungarian agriculture so that by 1980 its grain production had doubled that of the early 1960s, meat production was among the highest in Eastern Europe, and by 1973 Hungary was able to export over a million tons of grain annually. The Hungarian experience showed that market economics can work without private ownership of land if the producers effectively control their work.

Unfortunately, by late 1987 it became apparent that although Hungary still enjoyed one of the highest living standards in Eastern Europe, it was no longer the "economic miracle" of the Soviet bloc. Hungarian officials revealed for the first time in public that the nation's external debt had increased very rapidly in the last few years to $16 billion. Interest payments alone were more than $800 million per year. Hungary was having an increasingly difficult time repaying the debt because hard-currency export growth had not kept up with imports. To alleviate the economic problems, the government imposed price increases for basic commodities and proposed the introduction of value-added and personal income taxes. The prosperity contributed to the economy by the controlled dose of capitalism introduced in 1968 became an embarrassment to the Hungarian Socialist Worker's Party when private enterprise ventures became more efficient and successful than the state-run businesses. Too much success in private business was not tolerated. For example, the Pepi poultry enterprise, which was eight to ten times more profitable than the leading state-run poultry collective, was seized by the government. The owners, deprived of their investment and property, ended up as political refugees in the U.S.

For many years the economies of the Soviet Union and its eastern bloc allies have been mired in increasing stagnation. Their problems, rooted in their centrally planned systems, led to malaise which contributed to political upheavals and waves of democratization. Soviet president Mikhail Gorbachev's policy of *glasnost* and *perestroika* programs ignited the blaze that swept through all of Eastern Europe in late 1989 and early 1990. Positive results in terms of economic recovery will probably be slow and painful, but the new freedoms have stimulated some encouraging developments with regard to environmental protection. Taking advantage of the opportunities glasnost presented, citizens demanded government action to correct the environmental destruction brought about by the nation's ruinous industrial drive.

China presents another example of how central planning is not the answer to food production problems, does not forestall environmental destruction, or prevent the costs from being passed on to future generations. In the northern and northwestern parts of China there are 425,000 square miles of desert. More than half of this area is true climatic desert with less than four inches of rainfall annually, while 66,000 square miles is desert resulting from human abuse of the land by overcultivation, overgrazing, firewood collection, urbanization, and road building. Another 100,000 square miles of arid steppe is either undergoing desertification or is threatened by imminent desertification. Of the 66,000 square miles of humanmade desert, 46,000 square miles became deserts during the last 2,500 years, but the other 20,000 square miles resulted from human abuse within the last 50 years.

Environmental destruction existed in China before the communist takeover in 1949, but it intensified during the years of central planning. For example, during the 1960s Mao Tse-tung's Great Leap Forward planned tree-cutting in Sichuan Province to provide charcoal for backyard steel furnaces. Forest destruction exceeded forest replacement by 160 percent. This contributed considerably to the deforestation which is China's biggest environmental problem. Forest area in China declined by 25 percent between 1949 and 1980. During the same period the Maoist "grain first" or "grow more grain" campaign led to unsound expansion of cropland into marginal grasslands. This caused a serious decrease in cattle production and severe erosion. Mao's policies caused enormous losses in agricultural production and famine in some areas.

Mao's Great Leap Forward was followed by retrenchment and the violent Cultural Revolution (1966–1976) characterized by decentralized economic control, but with decision-making authority by local and provincial bureaucrats. Finally in December 1978 China abandoned the Soviet-type system and shifted to a market-oriented agricultural system. This stimulated a phenomenal surge in production. New pricing policies gave farmers incentive to increase their production. Within a few years these reforms led to an increase in agricultural output by more than 50 percent, an increase of rural incomes by two-thirds and a similar increase in industrial production. Between 1978 and 1986 grain output increased by one third, oilseed production doubled and output of meat rose 80 percent. Cultivated area was increased by only about 4 percent. Highly erodible land was fallowed and there was a dramatic increase in fertilizer use and employment of Green Revolution products. All this was accompanied by a decrease in the use of water and pesticides. Afforestation was encouraged by giving incentive to peasants to plant trees with the goal to extend forest over 20 percent of the land by the year 2000. Today China is the world's second largest food producer and achieved food self-sufficiency around 1980. The Chinese have acknowledged that unsound land use and the pressures of population growth were the primary causes of environmental deterioration. Their current reforms in agricultural practices and market-oriented economy have led to dramatic improvement in their overall situation. However, since 1984 China's grain production has fallen. Efforts to regain the 1984 level of harvest failed four years in a row, and in 1988 China imported 5 percent of its grain, about 15 million tons. With its dense population, China has a small cropland area per person, and considering its rapid rate of industrial development which pulls both labor and land away from agriculture, China may have to import a growing share of its food during the 1990s.

India contains some 600,000 square miles of arid and semiarid regions. Most of the country of Pakistan is arid or semiarid and the contiguous regions in the western part of the Indian Republic form the Thar Desert (about 172,200 square miles), the main arid region of the subcontinent. About 88 percent of the arid land in India is in the Thar Desert. The limits of the arid and semiarid regions of the subcontinent have not changed appreciably during the past century and there is little desertification in new areas. This stability is due largely to the meteorological influence of the Himalaya Mountains which protect central India from becoming more arid by controlling rainfall and feeding the rivers. However, further deterioration of soil conditions within the arid lands has been occurring. In both India and Pakistan salinization and waterlogging plague much of the irrigated land. For example, in the Mona reclamation area of Pakistan, about 90 percent of the soils were waterlogged by the early 1960s. Efforts are now underway to rehabilitate the stricken area, but such efforts are expensive and time-consuming. The requirements of 785 million people for food and fuelwood and of 260 million cattle and 120 million goats and sheep for grazing and browsing are inflicting serious destruction of vegetative cover. This has caused continual fresh deposition of sand and is more alarming than the actual expansion of arid and semiarid land.

Satellite data indicate that India lost some 16 percent of its forests between 1973 and 1981. The depletion of forests in areas within 60 miles of India's 41 largest cities has pushed fuelwood prices up as much as 42 percent in some areas. Flooding and silting of reservoirs are other direct results of deforestation. During the early 1980s damage from flooding in the Himalayan watershed averaged $250 million per year. Unfortunately, the continuing depletion of India's woodlands and continuing rise in fuelwood prices seem inevitable.

In the 1960s India, as well as most of Asia, was faced with a serious food crisis. Disaster was averted largely by the simultaneous development of new food plant varieties and improved agricultural techniques which greatly increased crop yields. This development, which resulted from research sponsored and supported by the Rockefeller and Ford foundations, became known as the Green Revolution. Plant breeders developed new varieties of wheat, rice, and other crops that were appropriate for the conditions in India, Mexico, Pakistan, Turkey, and other countries. In

India the new early-maturing varieties of wheat and rice along with an increase in irrigation to make dry season cropping possible, enabled many farmers to grow a summer rice crop and a winter wheat crop on the same land. Wheat production more than quadrupled between 1964 and 1984 when it reached 45 million tons. In India, as in China, wheat is beginning to replace rice as a favored food crop. Recently, as did China, India achieved food self-sufficiency. Whether India can sustain this achievement depends on how successful the nation is in reducing its annual population growth rate which persists at 2.3 percent, halting deforestation and soil erosion, and preventing waterlogging and salting of irrigated fields. The Green Revolution success increased output enough to eliminate grain imports, but did not increase per capita food consumption markedly. Its grain reserves enabled India to survive the poor monsoon of 1987, but another bad year would put the nation in serious trouble.

Among the smaller nations of Asia, there are wide variations in population growth rates, food production, and management of resources. Thailand, Singapore, and Indonesia have been able to reduce their population growth rates from high to moderate, but Burma, Vietnam, Malaysia, and the Philippines have made little progress in controlling population growth. These countries, as well as Bangladesh and Pakistan, benefited from the Green Revolution seeds. For example, Indonesia, formerly the world's largest importer of rice is now able to grow enough for its own needs and a surplus for export, but little or no progress has been made in expanding grain production since 1984. Burma, on the other hand, has suffered the reverse of this situation. Once able to feed most of Southeast Asia, Burma now is incapable of feeding itself. This appears to be the result of General Ne Win's brand of socialism. Faced with the specter of food shortages, the Ne Win government recently repealed laws that forced rice farmers to sell their crop to the state at low fixed prices, and now allows free-market dealing in rice.

Deforestation is a serious problem in Southeast Asia. The devastating floods in Bangladesh in 1988 were caused by the progressive deforestation of the Himalayan watershed. Tropical rain forests in Indonesia and Malaysia were degraded and destabilized by land-clearing programs, slash-and-burn farming and commercial logging. In 1982 and 1983 some seven forest fires spread through this region destroying over 8.6 million acres of tropical rain forest.

Conversion of the natural forests to rubber and palm oil plantations in Malaysia has doubled peak rainfall run-off and cut dry-season flows in half. In Sri Lanka the ambitious Mahaweli Development Program failed to achieve its goals of tripling the nation's electric generating capacity and irrigating an additional 320,000 acres of cropland by construction of large dams across the Mahaweli River. The failure was due to serious design and construction problems and deforestation of the hillsides causing runoff of soil downstream silting in the reservoirs and irrigation canals. Clearly, the problems in Asia are not as serious as those facing Africa, but the prospects for the future are not encouraging in the absence of appropriate corrective action.

Middle East

The Middle East is made up of the seventeen countries included in the area between Afghanistan and Egypt (since Egypt is mainly African, it was considered earlier). These countries occupy over 3 million square miles, nearly all of which is arid or semiarid. In spite of its deficiency in fresh water and rainfall, it is recognized as the cradle of civilization, and has from the beginning of recorded history provided humankind with important food crops including wheat, barley, chickpeas, figs, and pistachio nuts. Flax, rue, and the opium poppy originated in this region and many native plants offer the potential of becoming new crops that may provide food, forage, fibers, medicinals, and chemicals in the future. In the past, several of the Middle Eastern countries produced enough food to meet their own needs and a surplus for export. Today most of these nations depend on imports to meet their food needs, and petroleum has become the primary economic resource of the region.

Of the global oil reserves, 56 percent are in the Middle East. This compares with North America's share which is only 6 percent of the total reserves. The Middle East's share is actually rising because reserves in other parts of the world are being depleted more rapidly. The current estimated population of the Middle East, including Egypt, is about 178 million and is growing at a rate of 2.8 percent annually. Thus less than 4 percent of the world's people possess 56 percent of the world's remaining oil supply.

Distribution of population and resources is tremendously variable in the Middle East. For example, Saudi Arabia with a population of less than 9 million occupies an area of 1.4 million square miles and possesses more oil than any other country in the world, whereas Israel with an area of only 1,300 square miles has a population of over 4 million and virtually no petroleum reserves. In addition to Saudi Arabia, Iran, Iraq, Kuwait, the United Arab Emirates, Qatar, and Oman are oil-rich and depend heavily on oil exports. Syria, Lebanon, Turkey, and Jordan produce little or no oil, but receive income from pipelines passing through their land. Iran and Iraq depend on their income from oil exports, but the majority of the people are farmers. Per capita income also varies tremendously, from a high per capita figure of about $24,000 per year in Kuwait to a low of about $250 per year in Afghanistan.

The Middle East, which was the site of many early civilizations, has suffered from thousands of years of deforestation, overgrazing, and unwise cropping. Human activities have transformed large expanses into desert, and in spite of governmental efforts to thwart desertification, land degradation continues. In northern Iraq rangelands which could sustain 250,000 sheep safely are being forced to feed a million or more.

On Syrian rangeland there are three times the number of animals grazing than the land can safely support. Overgrazing first causes inferior plants to replace the more useful varieties, so that sheep pastures become suitable only for goats and camels. Finally, in the advanced stages of deterioration, plant cover disappears and the rangelands becomes sandy desert. This is happening in many of the steppe areas of Syria, Jordan, Iraq, and the United Arab Emirates. Furthermore, the 1991 Operation Desert Storm war caused significant additional economic, environmental, and social problems for Iraq and Kuwait.

Groundwater is utilized in many areas of the Middle East to supplement the meager and erratic rainfall to grow crops. Currently, over-pumping is causing falling water levels in the Arabian aquifer which underlies Saudi Arabia, Bahrain, Oman, Qatar, North and South Yemen, and the United Arab Emirates. In these oil-producing countries, salt-water is pumped into oil wells to raise oil pressure in some areas. This is introducing a salinity problem in some areas. Egypt's problems with the Aswan High Dam have already been mentioned. Food production has not been able to keep up with the demands of a steadily growing population. The decline in food output has been serious in Iraq, Jordan, and Lebanon. In Lebanon the problems are due mainly to complex political conflicts and bloody civil war. This tiny nation located on the Mediterranean Sea between Syria and Israel, was the homeland of the ancient Phoenicians who are credited with the invention of the phonetic alphabet and were the first to use the North Star for navigation. The landscape varies from seacoast to mountains up to 8,000 feet, from barren, rocky terrain to fertile valleys and plains in a country only 130 miles long and 46 miles wide. Until recently this was a land of beauty and sophistication with thriving agriculture and tourism.

Israel has been uncommonly successful in mobilizing people to halt deterioration of its arid land and restore it to a high level of productivity. The Negev Desert, which had been degraded by thousands of years of overgrazing and deforestation is today restored and productive after the introduction of innovative irrigation techniques, improved dryland farming, and carefully controlled grazing. This achievement provides hope that if peace can be restored to this troubled region, most of the degraded land has not passed the point of no return.

SELECTED INFORMATION SOURCES

Brown, L. R. 1982. *U.S. and Soviet Agriculture: The Shifting Balance of Power.* Worldwatch Paper 51. Washington, D.C.: Worldwatch Institute.

Brown, L. R., and E. C. Wolf. 1984. *Soil Erosion: Quiet Crisis in the World Economy.* Worldwatch Paper 60. Washington, D.C.: Worldwatch Institute.

Dahl, B. E., and C. M. McKell. 1986. "Use and Abuse of China's Deserts and Rangelands." *Journal of Range Management* 8:267–271.

Daly, J. J., and G. S. Dudgeon. 1987. "Drought Management Reduces Degradation." *Queensland Agriculture Journal* January-February:45–49.

Ellis, W. S., and D. C. Turnley. 1990. "The Aral—A Soviet Sea Lies Dying." *National Geographic* 177(2):73–93.

FAO Production Yearbook. 1989. Volume 43. Lanhan, Md.: UNIPUB.

Hughes, K. K. 1987. "Soil Salinity and Salting Can Be Prevented." *Queensland Agriculture Journal* January-February:27–30.

MacDonald, L. H. 1986. *Natural Resources Development in the Sahel: The Role of the United Nations System.* Tokyo: United Nations University.

Office of Technology Assessment, U.S. Congress. 1984. *Africa Tomorrow: Issues in Technology, Agriculture, and U.S. Foreign Aid—A Technical Memorandum.* Washington, D.C.: U.S. Government Printing Office.

Office of Technology Assessment, U.S. Congress. 1986. *Continuing the Commitment: Agricultural Development in the Sahel—Special Report.* Washington, D.C. U.S. Government Printing Office.

Pollack, M. 1987. "Eastern Europe—Stoking an Environmental Disaster." *Insight* September 7:38–39.

Pollack, M. 1987. "Hungary—Persistent Economic Strain Wears on People's Will." *Insight* November 23:28–30.

"USSR Promotes Personal Interest to Meet Higher Agricultural Standards." 1987. *Agriculture International* 39:4.

Walsh, J. 1986. "Crop Research Network Makes Some Changes." *Science* 234:1190–1191.

Walsh, J. 1986. "Grasshopper Control Program Successful." *Science* 234:815–816.

Walsh, J. 1986. "Return of the Locust: A Cloud Over Africa." *Science* 234:17–19.

TWO

The Plight of Agriculture in the Western Hemisphere

North America

The North American continent includes the United States, Canada, Greenland, Mexico, Central America, and the islands of the Caribbean. The total land mass is about 9,375,000 square miles, of which approximately 16 percent, or 1,500,000 square miles, is arid or semiarid with an average of less than twenty inches of rainfall per year. Most of this dry land is located in Mexico, the United States, and Canada west of the hundredth meridian and may be divided into ten regions as indicated in table 2.1.

The North American drylands constitute a great topographical variety, including mountains, hills and mesas, high plains and plateaus, precipitous canyons, and closed basins ranging in altitude from nearly 300 feet below sea level in the Mojave Desert to 11,000 feet above sea level in the Colorado Plateau. Rainfall in most of this area is extremely limited because of the rainshadow effect of high mountain ranges, and evaporation is generally very high. The diversity in vegetation is no less varied than the physiography, and the variation is as great with altitude as it is with latitude. Many of these plants, well adapted to arid and semiarid lands,

TABLE 2.1
North American Desert Regions

Region	*Size (sq. mi.)*	*Location*
1. Great Basin Desert	193,000	Most of Nevada, western Utah, northern Idaho, southeastern Oregon, northeast corner of California
2. Mojave Desert	25,000	Extreme southern Nevada, southeastern California, northwest corner of Arizona
3. Sonoran Desert	115,800	Southern Arizona, western Sonora, most of Baja California
4. Chihuahuan Desert	135,100	Mexican States of Chihuahua, Coahuila, San Luis Potosi, northwestern Nuevo Leon, western Texas, and southern New Mexico
5. Columbia-Snake River Plateau	100,000	Southern British Columbia, eastern Oregon and Washington, southwestern Idaho
6. California Valley	20,000	California between the Coast Ranges and the Sierra Nevadas, including the San Joaquin, Sacramento, and Tulare basins
7. Wyoming Basin	42,300	Southwestern Wyoming
8. Colorado Plateau	127,380	Western Colorado, southwestern Wyoming, eastern Utah, northeastern Arizona, and northwestern New Mexico
9. Great Plains	568,200	From Alberta, Canada to Mexican border between Rocky Mountains and 100th meridian
10. Southern Temperate Grassland	96,500	Southeastern Arizona and southwestern New Mexico southward into Mexico between Chihuahuan Desert and Sierra Madre Occidental

have potential for food, forage, fuel, chemicals, and medicines. Thus far they have not been exploited, mainly because of the circumstances of settlement and need. When the European settlers moved into the region they brought their own crop plants with them and, for the most part, ignored the potential offered by native plants used by the indigenous population. The land was made suitable for the introduced crops by clearing, tilling, irrigating, etc., rather than taking advantage of crops already suited to the land. The greatest limiting factor to the development of this region was, and still is, water. Many streams are ephemeral, and all of the large streams have been dammed to provide water for irrigation and hydroelectric power. In many areas salinity and dissolved solids are serious problems. Good quality water is often found near mountains as snow pack runoff in streams and in groundwater. Most of the land in the arid and semiarid regions of North America is suitable for grazing and some of it, when water is available, is important as cropland. Unfortunately, overgrazing and unwise farming practices have deteriorated many grasslands and markedly reduced their capacity for sustained productivity. Overdraft of groundwater, waterlogging, and salinization threaten the future of much of North America's most productive cropland. The regions where the threat is the most serious are the Great Plains and California's Central and Imperial valleys.

The Great Plains

Until the mid-nineteenth century this vast, nearly treeless flat grassland supported millions of buffalo and the predators which feasted on them. The latter included the nomadic Indian tribes: Sioux, Arapaho, Comanche, and Cheyenne, but no permanent civilization was ever established. By the 1860s and 1870s most of the bison and Indians were gone and great cattle drives from Texas to Kansas took their place. These resulted in drought, overgrazing, and falling meat prices. Following several wet years during and after World War I, the demand for bread encouraged the settlers to grow wheat on a grand scale. Increased mechanization, cheap energy, and rising market prices enabled farmers to put more and more land under cultivation. To keep up payments on their tractors and other implements, farmers could no longer afford to leave land fallow for a year. More and more marginal land previously needed for grazing, was put under the plough. Beginning in 1930 there were several years of drought, and millions of acres of wheat were so poor that the crop was not worth harvesting. Farmers could not meet their mortgage payments, and their livestock died.

In the fall of 1933 the dust storms started. The storms continued through the spring and summer of 1934. The clouds of dust obscured the sun as far east as New York and Washington. Three hundred and fifty million tons of dust, which had once been topsoil in the Great Plains, was scattered all the way to Europe. The National Resources Board reported that 35 million acres had been essentially destroyed and another 125 million acres were severely damaged. This was the infamous Dust Bowl of the 1930s. It was one of the worst humanmade catastrophes in history and it forced the westward migration of some three-quarters of a million bankrupted farmers, bankers, and merchants to California, Washington, and Oregon.

During these hard times one could still sink a well, erect a windmill-driven pump and get enough water for a family, some stock, and perhaps a garden, but not enough to irrigate large fields. Then came the centrifugal pump, which could raise 1,000 gallons per minute or more,and cheap energy in the form of gas or electricity to power it. When geologists studied the groundwater potential they found an immense aquifer underlying parts of eight states from southern South Dakota to the panhandle of north Texas—the Ogallala Aquifer. This closed-basin aquifer, representing trapped runoff of several ice ages, contained a volume of fresh water equal to that in Lake Huron, some three billion acre-feet. The thickness of the aquifer varied from more than 1,000 feet in parts of Nebraska to less than 50 feet along the periphery in parts of Texas. This discovery, coupled with the perfecting of deep-well-drilling techniques, the invention of efficient deep-well pumps and center-pivot irrigation systems, and the availability of electricity and natural gas for power, transformed the Great Plains from dryland farming to irrigated farming within a few short years. The water was free, but the capital investment in a standard (130–135 acre) center-pivot system including the cost of a deep well, a deep-well pump, and necessary pipeline could easily exceed $50,000, a high figure at the time.

The Ogallala Aquifer is the only reliable source of water for most of the Great Plains farms, ranches, and towns. The area of irrigated farmland grew from 2 million acres in 1949 to over 13 million acres in 1980. Twenty percent of the irrigated cropland in the United States is supported by water mined from the Ogallala Aquifer. In 1977 this irrigation made it possible for the farmers of the region to produce $2 billion worth of crops and $10 billion worth of beef, 40 percent of the nation's supply. One of the poorest farming areas in the United States had suddenly become one of the richest. The entire economy benefited; in addition to a significant increase in farmers'income, allied businesses such as banks, machinery, fertilizer, seed, and fuel dealers, grain elevators, meatpackers, and trucking services all flourished. Local, state, and federal governments benefited from increased tax revenues. Cities like Lubbock and Amarillo, Texas grew at rates of 7.5 percent per annum, and skyscrapers rose from the plains.

The Ogallala Aquifer supports a mining industry dependent on water deposits millions of years old. Recharge of the aquifer from precipitation is almost negligible, ranging from a rate of less than half an inch per year in parts of New Mexico to nearly six inches under the Sand Hills of Kansas and Nebraska. In 1984 hydrologists of the U.S. Geological Survey estimated that the aquifer was half depleted under 2,224,000 acres of Kansas, New Mexico, and Texas. In recent years farmers in these areas, faced with rising pumping costs, diminishing water yields, and lower commodity prices, have been forced to reduce the amount of land under irrigation. Between 1978 and 1982 the area of irrigated land dropped 20 percent in Texas, 18 percent in Oklahoma, and 9 percent in New Mexico. The total irrigated area in the region declined by some 1,463,000 acres or 7 percent. At the same time, irrigated acreage was being expanded in Nebraska where a smaller portion of the aquifer had been depleted. By 1982 it was generally recognized that groundwater depletion of the Ogallala Aquifer posed a serious problem for both the region and the nation.

Publicity on this impending disaster stimulated Congress to authorize and fund a number of federal studies, the most important of which was the 1982 Six-State High Plains-Ogallala Area Study coordinated by the Economic Development Administration of the Department of Commerce. The study concluded that serious local water supply problems exist and that they in turn cause serious economic problems, but that the entire region was not in immediate danger. In the long term, however, disaster was predicted, starting in the southern plains. By the year 2020 the Texas share of the Ogallala would be only about 30 percent of the 283.7 million acre-feet it had in 1977,and the New Mexico share would be all but used up. Colorado and Kansas would be somewhat better off, and Nebraska would still have 1.9 billion acre-feet and could support irrigation of 11.5 million acres, more than any other state in the union. Therefore, irrigation farming would move northward and Texas, New Mexico, and Oklahoma would return to dry-land farming. The study considered possibilities for augmenting the water supply to the region by diverting Missouri River water and water from the Arkansas-Red-White River system, but concluded that interstate supply augmentation would not be a viable option.

The California Central and Imperial Valleys

The California Central Valley is more than 400 miles long, stretching from Redding on the north to Bakersfield at the southern end. This valley lies between the coastal ranges on the west and the high Sierra Nevada Mountains on the east, and includes three large basins: the Sacramento, San Joaquin, and Tulare. Elevations range from about 400 to 1,600 feet with the drainage from the Sacramento and San Joaquin rivers into the Pacific Ocean via San Francisco Bay. The southern end of the valley is a closed basin with drainage into playas known as Tulare and Bueno Vista lakes. In the southeasternmost part of California lies the region referred to as the Imperial Valley. Average annual rainfall is generally less than ten inches and it occurs mainly during the mild winters. Irrigation from groundwater supplies and surface diversions have made it possible for the California valleys to contribute a sizable share of the world's food production. Unfortunately, huge demands on the groundwater reserves lowered the water table and caused settling of the land. Serious salt and soil saturation problems also plague these valleys.

The San Joaquin Valley, located in the southern two-thirds of the Central Valley of California, is an important agricultural region with a long history of salinity problems. Irrigated acreage grew rapidly from the late 1870s to about 1915. In those days most of the water used for irrigation was groundwater. By 1930 this source was nearly exhausted and the state's biggest industry was threatened with collapse. The farmers pressured the state and federal governments to authorize the California Water Project and the Central Valley Project, together amounting to the largest water project in the world. The two projects have collected enough water to supply eight cities the size of New York.

In a complex system of dams, reservoirs, canals, tunnels, and aqueducts the projects have captured the water from the San Joaquin River, which drains the southern half of the Sierra Nevada Mountains, and also the flow of the Sacramento River which drains the northern half of the Sierra Nevada and some of the coast range. Some two-thirds of the runoff from the nation's third largest state is collected and moved from one end of the state to the other. The natural flow of the Sacramento watershed south and the San Joaquin watershed north flows into the delta just southwest of the city of Sacramento and from there into San Francisco Bay and the Pacific Ocean. Water from the Sacramento River and its large tributary, the Feather River, is shunted past the delta and pumped up 300 feet to enter the California Aqueduct. This humanmade "river" is 444 miles long. It parallels Interstate Highway 5 for 250 miles through the San Joaquin Valley and provides irrigation water for rich farms and orchards. Fifty miles east of Santa Cruz the water is interrupted at the San Luis Dam, forming a huge reservoir that provides stability and security against unpredictable weather conditions. At the south end of the San Joaquin Valley the aqueduct encounters the Tehachapi Range where the water is lifted 1,926 feet to bring it to Los Angeles. The Edmonston pumps which lift this sizable river uphill consume 6 billion kilowatt-hours of electricity per year—more electrical energy than is used by several states for all electrical needs. Some electrical energy is gained back when the water plunges downhill on the south side of the Tehachapi Range through a battery of turbines. Below this point it becomes an open aqueduct again, a concrete highway of water with forks that sends the west branch to Los Angeles and the east branch across the Mojave Desert toward Riverside and ultimately to Lake Perris.

Altogether there are some 1,200 dams and reservoirs in California; these with their thousands of canals, aqueducts, and tunnels along with deep wells pumping groundwater as if it would last forever, provide some 34.2 million acre-feet of water for 28 million people, thriving industries, and a $15 billion agricultural industry. Large as the domestic industrial and municipal water needs are in this arid region, they are dwarfed by the quantities consumed by the farmers: 83 percent of the total is used in agriculture, and most of the enormous cost to bring this water to southern California from the northern part of the state and from the Colorado River is paid by the taxpayer. California in 1991 is facing its fifth consecutive year of drought, and every year the population of southern California increases by about 350,000. Urbanization of farmland is spreading at a rapid pace. In the Central Valley nearly 20,000 acres of farmland is being converted to urban use every year. Fresno is the fastest-growing big city in the U.S. with a population increase of 61 percent since 1980. Urban areas are faced with mandatory cutbacks in water supply, while farmers are largely unrestricted in the use of flood irrigation with heavily subsidized water. The competition between the booming cities and the powerful agricultural community for the scarce water is mounting ominously. Farmers are charged only $2.50 to $19.31 per acre-foot of water from the Central Valley Project, while the Metropolitan Water District has to pay $233 an acre-foot to supply water to urban southern California. The story of the relentless quest for water in the American West is vividly detailed in a 1986 book by Marc Reisner entitled *Cadillac*

Desert, The American Desert and its Disappearing Water. On a grand scale, federally subsidized water is used to irrigate land to grow surplus crops, their prices are propped up by federal subsidies, and the surpluses are stored at government expense.

The San Joaquin Valley faces another, even worse, problem. Water, made available to this valley at an awesome cost, transformed it into the richest agricultural region in the world. But underneath some two million acres of this fabulously productive irrigated desert is a layer of clay through which little or no water can pass. In the middle of the valley this nearly impenetrable layer of clay is located just a few feet under the surface soil. Irrigation water collects on top of the clay and is termed "perched" water by hydrologists. In the dry heat of the valley the evaporation rate is high and the perched water cannot penetrate through to the fresh water aquifer under the clay thereby increasing the salt concentration steadily. The more the farmers irrigate, the worse the waterlogging and salinization gets; already thousands of acres near the southern end of the valley are all but destroyed.

To solve this problem, the state and federal planners considered the need for a master drain to carry the perched water out of the valley. For thirty years various schemes have been proposed, but to date no overall plan has been approved. The Bureau of Reclamation proceeded alone in 1968 to construct the federal San Luis Drain, a multi-million-dollar facility that was supposed to extend from Kettleman City 188 miles north to the delta. To date only an 82-mile segment of the drain has been completed from southern Fresno County to a humanmade swamp called Kesterson Reservoir just north of Los Banos. The reservoir attracted thousands of migratory waterfowl because over 90 percent of the original wetlands of the Central Valley have dried up as a result of the action of the Bureau of Reclamation. The presence of all the ducks, geese, shore birds, and other species stimulated the bureau to write off a portion of the enormous cost of the drain as a wildlife and recreational benefit. The marsh also provides a home for many species of fish, amphibians, and small mammals.

In 1982 the wardens at the Kesterson Reservoir National Wildlife Refuge noted that something was seriously wrong when they found that all but one of the fish species had disappeared, and the next spring many ducks were born dead or horribly deformed. Biologists soon determined the most likely culprit to be high concentrations of selenium in the drainage water—200 to 400 parts per billion—washing down from the southern Coast Range soils. Selenium is an element essential for human nutrition, but in very tiny quantities. The amounts found in the Kesterson Reservoir were highly toxic. In addition to selenium, waste water contains high concentrations of boron and pesticides. In late 1984 the Kesterson Reservoir, the poisoned humanmade marsh, was declared a toxic waste dump.

Completion of the master drain faces two obstacles, which now appear to be insurmountable. First is the enormous cost. By the time the San Luis Drain portion of the master drain is completed the cost will be more than $500 million. In 1984, Secretary of the Interior William Clark projected that solving the drainage problem in the entire valley would cost $4 to $5 billion. If these figures are realistic, it means that it would cost about $5,000 per acre to save the affected land, more than the land is worth. Neither the farmers, wealthy as many of them are, nor the taxpayers are willing to foot the bill, however. The second obstacle is that the only ultimate destination for the waste to be carried by the master drain is San Francisco Bay. Now that the horrors of the Kesterson Reservoir disaster are well known, the 5 million people in the Bay Area and environmentalists everywhere are understandably reluctant to have the drain completed.

Before settlement by the white immigrants, the southeast corner of California, known as Imperial Valley, was the desolate Colorado Desert. Water from the Colorado River became available for irrigation in 1901 and by 1918 some 360,000 acres were irrigated and under cultivation. Because of the heavy soils and the salty Colorado River water, salinization had already by that time forced 50,000 acres out of production and damaged an additional 125,000 acres. A partial solution to the problem was found in the 1940s when the U.S. Soil Conservation Service in a joint program with the Imperial Irrigation District designed subsurface tile systems tailored to meet the drainage needs of individual landowners. The Salton Sea, 273 feet below sea level and nearly dry in 1904, was, and still is, the receptacle for the saline waste-water collected from the tile lines and channeled by gravity to the low-lying lake.

Since 1949 the valley's salt balance has been maintained and its large farms have flourished, but the irrigators have had to spend some $40.5 million for drains between 1929 and 1972, and over $26.2 million on concrete lining for canals and laterals between 1954 and 1972. The rising salinity levels of the Colorado River require increasingly sophisticated irrigation and soil mana techniques. If the salinity levels of the river reach the 1,140 t parts per million range projected for the year 2000, reduced and increasing operating costs could lead to significant eco problems for these farmers. Another set of problems centers around the Salton Sea. Increased runoff from expanded irrigation developments and more efficient drainage systems have caused the water level of the Salton Sea to rise nearly 40 feet during the last 40 years. This has caused millions of dollars in damage to shoreline properties. Because of the high evaporation rate and the deposit of about 5 million tons of salt per year, the salt concentration of the Salton Sea has risen to about 40,000 parts per million, 5,000 parts per million higher than ocean water. This level of salinity threatens to kill the salt-water fish and the sport fishing industry successfully introduced during the 1950s.

marana

The Sonoran Desert

Within this desert region are the cities of Phoenix and Tucson in Arizona, Mexicali in the Mexican state of Baja California Norte and Hermosillo and Ciudad Obregon in the Mexican state of Sonora. In addition to these population centers there are several significant agricultural areas in the region, and all are heavily dependent on irrigation water. The climate is brutally hot and dry. Rainfall is unpredictable and skimpy at best. Streams for the most part are ephemeral. The only reliable source of fresh water is groundwater, and both municipalities and farmers have been using it at rates that threaten to deplete the aquifers.

The last several decades have been boom years for Arizona. The population of Phoenix grew from 65,000 in 1940 to 439,000 in 1960, to 1.5 million for the metropolitan area in 1986. The state's population doubled twice between 1920 and 1960, and millions of acres of irrigated cropland came into production. Four-fifths of all water needed to support this growth has been groundwater in spite of the fact that the Bureau of Reclamation has captured nearly all of the Gila drainage in their Salt River Project. The annual overdraft has been in excess of 2.2 million acre-feet. This severe groundwater depletion has caused aquifers to collapse, producing subsidence and huge fissures in the desert east of Phoenix and problems on Interstate 10 between Phoenix and Tucson.

The city of Tucson with a population of nearly a half million is completely dependent on groundwater. This community is drawing groundwater out of neighboring basins now because it has already overdrawn its own. The Colorado River forms the boundary between Arizona and California, and represents a potential source of surface water for Phoenix, Tucson, and the agricultural land between them. After twenty-five years or more of fighting, negotiating, and politicking, it appears now that Arizona will get a portion of the Colorado River water under the Central Arizona Project (CAP). California is the big winner and is guaranteed 4.4 million acre-feet of water per year before Arizona gets a drop. The upper basin states, Wyoming, Colorado, and Utah, have not been using their full allotment of Colorado River water, and when they do there is a serious question of how much, if any, of the flow will be left for Arizona.

In addition to the domestic allotments, Mexico, by a 1944 treaty, was granted 1.5 million acre feet of Colorado River water each year. As a result, the Mexicali region is the most productive farmland in all of Mexico, but it is completely dependent on Colorado River water for irrigation. There were no serious problems in this area until the Bureau of Reclamation revived the Wellton-Mohawk Irrigation District along the lower Gila River Valley. The irrigation district receives its water from the Colorado River. This region is located above a salt dome and is plagued with poor drainage, so the bureau built, at considerable expense, an elaborate drainage system to carry the salt-laden waste water away through perforated tiles, which led into a master drain that emptied into the Colorado River just above the Mexican border. This project was completed in the early 1960s at the time the gates were closed on the Glen Canyon Dam. The combined effect of the influx of salt-laden water from the drain and the drastic reduction of fresher flows from upstream caused the salinity of the Colorado River to almost double from about 800 to more than 1,500 parts per million. Crop yields in Mexico's best agricultural region plummeted and the Mexicans were furious.

Luis Echeverria won the presidential election in 1973 largely because he promised to haul the United States before the World Court on this issue. In 1973, after the first OPEC oil crunch and data indicating that Mexico owned a lot of oil, President Richard Nixon appointed Herbert Brownell to work out a hasty solution. In August 1974 an agreement was signed for the United States to deliver to Mexico water with a salt content of not more than 115 parts per million (plus or minus 30 parts per million) higher than the level at Imperial Dam in 1976, or 879 parts per million.

The simplest and least expensive way for the U.S. to avoid violation of this international agreement would have been for the government to buy out the Wellton-Mohawk farmers for a few hundred thousand dollars and retire their lands. Instead, Congress approved the construction of the world's largest reverse-osmosis desalination plant at Yuma, Arizona to treat the waste water running out of the drain canal. According to Tom Turner of the *Arizona Daily Star*, by October 1986, only a pilot plant had been constructed at a cost of well over $200 million. Completion of the plant was scheduled for 1989 when it was expected to purify 72.4 million gallons of waste water per day or 80,000 acre feet per year. In addition to the desalination plant, the treaty with Mexico called for construction of a concrete-lined water drain to carry the brine waste from the desalination plant to the Gulf of California at a cost of $23 million, sealing a portion of the old California irrigation canal for $49 million, sinking a field of "counter wells" along the international boundary for $20 million, and removal of 25,000 acres of U.S. farmland from production for an undisclosed amount. Thanks to better than average rainfall and unusually heavy snows in the Rocky Mountains, the Colorado River has been at its highest level in forty years. This has made it possible for the U.S. to meet its treaty obligation to Mexico even though the desalting plant in Yuma is not operating. It also made it possible for Phoenix to receive Colorado River water via the CAP aqueduct in 1985. And if the favorable precipitation conditions hold, Tucson is scheduled to receive CAP water in 1992. Who will pay what for this water is still an unanswered question, but it seems likely that unless Congress is very generous, the water will cost the area farmers more than they can afford.

Arizona, and Tucson in particular, have taken important steps to conserve their meager water supplies. In 1980 the Arizona Legislature passed the Groundwater Management Act requiring all municipalities, irrigation districts, farmers, and industries to conserve water. The southern Arizona Water Resources Association has developed plans and guidelines to help homeowners, gardeners, farmers, and recreational facilities to conserve by landscaping with native desert plants, utilizing nonpotable "graywater" whenever possible, converting to drip irrigation systems, etc. The people of the region are beginning to understand the need for water conservation, and well they might before they suffer the fate of the Hohokam civilization which flourished in the Salt River Valley near present-day Phoenix from 200 B.C.to 1400 A.D. The Hohokam civilization disappeared during a drought that tree-ring data indicate lasted some fifty years.

Depletion of groundwater and salt accumulation in streams and irrigation drains is not limited to the region so far discussed. Similar problems are being encountered in many areas. El Paso, Texas and Mexico City are faced with serious groundwater overdrafts and so are parts of eastern and central Oregon. In parts of Mexico near the coast and in cities like Hermosillo and Ciudad Obregon the overdraft is so serious that seawater is invading the aquifers. The cost of wind and water erosion, salinity, and acidification on the Canadian prairies has been calculated at more than 600 million Canadian dollars a year. In New Mexico, 30 miles downstream from a heavily irrigated strip of land along the Pecos River, salinity levels have risen to 2,020 parts per million. The Arkansas River is suffering a similar plight. These problems are not all in the West. Long Island, New York is depleting its closed-basin aquifer and poisoning it with chemical wastes.

South America

South America, with an area of 6,864,000 square miles, lies mainly in the southern hemisphere and is divided politically into thirteen nations: Argentina, Bolivia, Brazil, Chile, Colombia, Ecuador, French Guiana, Guyana, Paraguay, Peru, Suriname, Uruguay, and Venezuela. The equator crosses through the broad northern part of the continent producing a torrid zone. The continent tapers off to a point at the southern end where the climate is subantarctic. The lofty Andes Mountains form the backbone of the continent, traversing along the western side from the Caribbean to the archipelago of southern Chile. In the central portion of the continent in Peru, Ecuador, Bolivia, and the northwest part of Argentina the Andes widen into complex and transverse ranges creating a broad band of highland and mountain country. To the south the Andes form a simple double chain. In the north, to the east of the Andes, is the jungle land of the Amazon River, cut off by the highland of Guiana from the hot coastal region and the valleys of the Orinoco River in Venezuela and Magdalena River in Colombia. South of the enormous Amazon watershed are the tablelands of Matto Grosso and the rich plateau of central and southern Brazil. Uruguay and northern Argentina are dominated by plains country pampas which give way to the bleak and semi-arid steppes of Patagonia. West of the Andes from Patagonia are the heavily forested southern islands of Chile, and to the north of the islands is the Mediterranean-like portion of central Chile. Still farther north in Chile is the Atacama Desert, a pitilessly arid nitrate desert which continues northward along the coasts of Peru and Ecuador. The coastal areas of northern Colombia and Venezuela next to the Caribbean Sea are hot and semiarid with less than 20 inches of rainfall annually. The other semiarid region of South America is in northeastern Brazil where precipitation averages between 15 and 28 inches annually with wide variability and a mean temperature of over 80 degrees F. The problem areas of South America include virtually all of the arid and semiarid lands. In the arid Coquimbo region of Chile, which lies between the lifeless Atacama Desert to the north and the irrigated valleys of central Chile, overgrazing has led to serious damage to the land. Cacti replaced shrubs in some locations, and in other areas native perennial plants have given way to less productive annuals and weeds. With the decline in the quality of the range, cattle are displaced by sheep and then goats replace the sheep. The inequitable land-tenure pattern in this region has promoted ecologically unsound agricultural practices. The wealthy landowners have large, sparsely populated estates which permit proper grazing rotations and acceptable agricultural practices, but the land holdings of the poor majority are overcrowded and increasingly degraded. Cultivation of hillsides has led to massive soil erosion and total loss of topsoil on some of the steeper slopes. Desertification is a serious problem in this part of South America.

In Argentina a much larger area is threatened. In the states of La Rioja, San Luis, and La Pampa desert-like environments are being created. In the irrigated sections of Argentina, a United Nations study has reported that salinization and alkalinization have reduced productivity on nearly 5 million acres. The same UN survey reported that in Peru nearly 40 percent of the 2 million acres of irrigated agricultural land in the coastal desert are affected by poor drainage and salinization. In Colombia a UN study indicated that the country loses 426 million tons of fertile topsoil each year. This country's topsoil is relatively thin; considering the poorly nourished state of many of its people, they can ill afford such a loss of resources. Soil erosion is a problem in the Peruvian Andes, causing declines in potato harvests. In some parts of Venezuela, growing population pressure has forced the traditional fallow period to become increasingly shorter, leading to decline in soil fertility and decreasing organic content of the soil. The water-holding capacity of the soil declines, soil structure deteriorates, and compaction becomes a problem. Thus, desertification has become a moderate hazard in this northern region of the continent.

Brazil, the largest country in Latin America, has two areas where environmental deterioration is especially alarming. The first area of concern is in the semiarid tip of northeastern Brazil where desert-like zones are expanding into the more humid interior because of massive deforestation and unwise agricultural practices of ranching corporations and farmers. The threat of desertification is very high to moderate over a wide area.

The other endangered area is in the state of Rondônia in the Amazon rainforests of northwestern Brazil. Here the threat is not only to the soil of Brazil but also of international concern. In 1982 the government of Brazil, with financing from the World Bank, initiated the Polonoroeste Project. Almost half of the nearly $500 million received from the World Bank was spent to build and pave Highway 364 to enable tens of thousands of migrants to move to the state of Rondônia. Between 1978 and 1984 some 400,000 settlers came to Rondônia only to find that the mineral-poor soil could not support cash crops. Most of the migrants were poor, landless peasants lured to the area by the prospect of free federal land. Thousands of these people cleared the tropical forest mainly by slash-and-burn methods. They planted crops and watched them fail because the infertile tropical rain forest soil is not suited to the agriculture they knew from their temperate-zone homes. Most of the migrants have been forced to sell their land to cattle ranchers or speculators and then work for subsistence wages. Some move on to forest reserves or Indian lands to repeat the destructive process. A photograph taken from the space shuttle showed an area half the size of the state of California engulfed in thousands of fires. This has led to an ever-widening circle of deforestation, and the destruction of more than 7 million acres of irreplaceable rain forest. According to recent calculations, if the destruction continues at the present rate, the state of Rondônia—the size of western Germany—will be denuded within a few years.

The cost of this well-intentioned but misguided agricultural development project has been enormous. The ancestral home of dozens of Brazilian Indian tribes, the habitat of thousands of plant and animal species—many of them rare or endangered and often unidentified scientifically—the wintering grounds of North American songbirds, all are being systematically destroyed by fire. The burning of the tropical forest is adding huge amounts of carbon dioxide to the atmosphere, contributing to the so-called greenhouse effect. All of these losses are taking place in an effort to clear rain forests whose soils are too infertile to sustain farming.

Under pressure from the Environmental Defense Fund (EDF), the Sierra Club, the Reagan administration, and Congress, the World Bank withdrew temporarily its support for part of the Polonoroeste Project in 1985 because of the environmental and social problems. Disbursements for road building in the Amazon rain forest have been interrupted until environmental concerns can be satisfied. Both the World Bank and the Inter-American Development Bank have endorsed the EDF recommendation to set aside large areas of Brazilian rain forest as "extractive reserves" where rubber tappers, nut gatherers, and other local people can harvest renewable forest products.

This is by no means a complete account of all the problems the world faces in providing food, shelter, clean water, and air for its present and future populations. But it is enough background to alert the reader to the seriousness of the dilemma and the need for corrective action. The following chapters are intended to show how this disaster can be turned around, that knowledge is available to solve most, if not all, of the problems, but human apathy and political inertia must be overcome to prevent wholesale catastrophe.

SELECTED INFORMATION SOURCES

Backlund, V. L., and R.R.Hoppes. 1984. "Status of Soil Salinity in California." *California Agriculture* 38:8–9.

Beck, L. A. 1984. "Case History: San Joaquin Valley." *California Agriculture* 38:13–16.

Bittinger, M. W., and E. B. Green. 1980. *You Never Miss the Water Till—(The Ogallala Story).* Chelsea, Mich.: Water Resources Publications.

Drylands Project. 1987. *Drylands Dilemma, A Solution to the Problem.* Executive Report, Drylands Project.

El-Ashry, M. T., and D. C. Gibbons. 1986. *Troubled Waters—New Policies for Managing Water in the American West.* Study no. 6. Washington, D.C.: World Resources Institute.

Ellis, W. S. 1991. "California's Harvest of Change." *National Geographic* 179(2):48–73.
Farrell, D. B. 1987. "Water and the Desert Dweller." *Arizona Highways* 63:4–11.
Goodin, J. R., and D. K. Northington, eds. 1985. *Plant Resources of Arid and Semiarid Lands—A Global Perspective*. Orlando, Fla.: Academic Press.
Holburt, M. B. 1984. "The Lower Colorado—A Salty River." *California Agriculture* 38:6–8.
Kelley, R. L., and R. L. Nye. 1984. "Historical Perspective on Salinity and Drainage Problems in California." *California Agriculture* 38:4–6.
Meyer, J. L., and J. van Schilfgaarde. 1984. "Case History: Salton Basin." *California Agriculture* 38:13–16.
Postel, S. 1984. *Water: Rethinking Management in an age of Scarcity.* Worldwatch Paper 62. Washington, D.C.: Worldwatch Institute.
Powers, W. L. 1987. "The Ogallala's Bounty Evaporates." *Science of Food and Agriculture* 5:2–5.
Ragan, M. L. 1991. "Amid Big Water Fight, California is Still Dry." *Insight* 7(2):20–22.
Reisner, M. 1986. *Cadillac Desert—The American West and Its Disappearing Water.* New York: Viking.
Sierra Club. 1986. *Bankrolling Disasters: International Development Banks and the Global Environment.* Brazil: Sierra Club, Ponoroeste Project.
Supalla, R. L., R. R. Lansford, and N. R. Gollehan. 1982. "Is the Ogallala Going Dry?" *Journal of Soil and Water Conservation* 37:310–314.
Turner, T. 1986. "Yuma Desalter to Purify Water for Mexicans." *Arizona Daily Star,* October 19.

THREE

Agricultural Practices in Arid Lands

The most important limiting factor in agricultural production in arid and semiarid lands is water. Protection of the land from erosion and maintenance of fertility are critical also. This chapter is devoted to the agricultural practices or methods available for dealing with these factors. It must be clear from the preceding chapters that: 1) physiography and climatic conditions vary widely in different parts of the world, and 2) agricultural practices have often led to deterioration of the land and desertification. The objective of this chapter is to describe useful practices and emerging technologies which provide improved efficiency, economy, and land preservation.

Irrigation Techniques

Construction of huge dams to provide water for irrigation is seen by many as an effective way to solve the water needs in arid lands. We have seen, however, that this approach is not without serious problems: the enormous cost of large dam construction, loss of cropland from inundation by the impoundments, silt collection limiting the useful life of the dam, salinization of irrigated farmland, and the need to construct expensive drainage systems. These problems have been encountered wherever large-scale dam-irrigation projects have been located. Today Africa is in desperate need to increase food production, and it has been proposed that irrigation could be a significant contributor to solving the food crisis. But a review of large-scale dam-irrigation projects indicates a gross failure to meet food production targets. For example, in Nigeria massive investments were made in these large-scale projects over a fifteen-year period. It was expected that some 700,000 acres of land would be brought under irrigation, but only about 10 percent of the goal was achieved. In Mali, Senegal, Burkina Faso, and other countries similar examples could be cited. The scale of the projects was often too large and the approach was "top-down" with management treating the small landholders as laborers. The farmers were told what to grow and where to grow it with no regard for local needs or experience. More often than not, low prices were paid to the farmers for their produce, offering no encouragement for efficient production. Because of the high cost and the experience of failure, most major donors have stopped lending for new large-scale irrigation projects.

Small-Scale Irrigation and Watershed Management

Attention is turning now to smaller-scale, informal, "bottom-up" irrigation approaches practiced by individual farmers requiring little or no capital investment, utilizing traditional and familiar methods. In contrast to the dismal record of the large-scale, formal irrigation projects in Nigeria, small-scale irrigation has grown from less than 300,000 acres to about 2 million acres in the last 25 years. Similar developments have been reported in the Ivory Coast, Liberia, Senegal, and Sierra Leone.

These successful ventures are modest in that they do not require large inputs of imported materials or machinery. They develop small local sources of water such as wells, streams, or moist valley bottoms rather than involving vast catchment areas. The projects are operated and managed by local farmers. They build on and improve what the farmers were doing, rather than forcing them to change to unfamiliar approaches. One of the objectives of these approaches is to increase the proportion of rainwater that filters into the soil. Any technique which maximizes infiltration protects against drought by replenishing the soil's moisture, raising the level of the water table, and refilling aquifers. Moreover, rainwater does not contain the burden of salts usually associated with irrigation water.

There are a number of irrigation techniques other than the familiar use of water from impoundments or utilization of diversion canals from rivers. Rainwater harvesting, water-spreading, microcatchment farming, and runoff agriculture have been practiced in arid lands for centuries and offer worthwhile possibilities for many areas today. Various schemes have been devised for collecting rainwater from hillsides or humanmade catchments. One technique, originally developed in the arid Negev Desert, concentrates rain falling on a wide area onto a smaller area where crops are grown. This method, illustrated in figure 3.1, not only channels and concentrates rainfall onto cropped fields, but also protects the area from erosion by allowing the water to spill gradually down the slope and soak into the soil rather than washing it away. Modifications of this technique are being employed to advantage in the arid and semiarid regions of Africa today. Many of them are described in detail in the excellent book by Paul Harrison entitled *The Greening of Africa. Breaking Through in the Battle for Land and Food.*

One African method described by Harrison involves nothing more complicated than lines of stones arranged along the contour to slow down the destructive force of runoff. This effective approach has been developed by Oxfam in the Yatenga area of Burkino Faso. These simple stone lines increase infiltration, improve crop yields, reduce erosion, and even rehabilitate thoroughly degraded land. In the Yatenga region the slopes are very slight, so it is often difficult to tell if the ground is sloping up or down. Since the stone lines must follow the contour to be effective, a simple, inexpensive device called a hosepipe water level is used to solve the problem. It consists of two stakes, about five feet long, marked halfway up with a series of lines about a tenth of an inch apart. A small-gauge transparent hosepipe, 30 to 60 feet long, is partially filled with water and one end is attached to each stake. When the bottoms of the two stakes are at the same level, the water in each end of the hose will rise to the same mark on each stake. Pickets are hammered into the soil to mark out the contour lines indicating where the stone lines should be placed. Illiterate peasants can easily master this technique and pass it on to their neighbors. The stone lines take up only 3–6 percent of the surface area. The con-

Fig. 3.1. Sketch of water-spreading dikes built in Pakistan. Zigzag pattern slows the torrent of floodwater and allows it to penetrate the soil. Crops are then planted in the wetted areas behind the dikes. (Adapted from French and Hussain, *Water Spreading Manual.* Range Management Record no. 1, West Pakistan Range Improvment Scheme, Lahore, Pakistan, 1964). Courtesy of National Academy of Sciences.

struction of the stone lines can be done during the dry season, so it does not conflict with food production. Most farmers try the method on barren land first; when they achieve significant yield increases even in the first year, they are impressed and the good news spreads rapidly. By the end of 1984 nearly all of the farmers in the area had built stone lines on at least some part of their land and were reaping decent harvests of sorghum and groundnuts.

In the American Midwest and High Plains thousands of farm and ranch ponds have been made by bulldozing a depression in strategic locations in a slope or hillside and piling the removed soil on the downhill side to form a dam. This form of rainwater harvesting is used mostly for watering livestock, but the ponds are used also for raising fish and, in some cases, provide irrigation water on a small scale for crops. In addition, water infiltration from the ponds helps to recharge underground aquifers.

Irrigation Systems in Developed Countries

Water for irrigation in most of the United States and other developed nations is obtained from reservoirs, streams, or groundwater and is distributed to farms via open channel canals (ditches) or pipelines. Application methods at the farm or ranch include surface, sprinkler, and drip irrigation. *Surface irrigation* is the oldest method; it utilizes the field to convey water through the area being irrigated by gravity flow. Furrows and dikes of various design may be used to control and direct the flow. Uniform distribution of water is often difficult to achieve. Land leveling using lasers is one modern solution. Laser land leveling is done with earth-moving machinery automatically controlled by a laser grade control system. This is an expensive operation, but can, in the long run, be worth the cost.

Sprinkler irrigation systems deliver water through the air from nozzles. This method allows more control over distribution of irrigation water than does surface irrigation, but evaporation losses are high. Several kinds of sprinkler systems are currently in use including permanent, solid set, hand move, and mechanical move. In recent years the center-pivot, mechanical move system has become widely used throughout the Great Plains and the Corn Belt. This system requires increased energy input, but has low labor requirement and is well adapted to steep terrain, sandy soils, and low water supplies. Research on sprinkler irrigation has focused on increasing application and distribution efficiency and reducing fuel requirements. One system, developed by William M. Lyle and co-workers, is the low-energy precision application (LEPA) system, which distributes water through drop tubes and low-pressure emitters directly into furrows as it moves in either linear or circular fashion through the field. It applies water uniformly with little evaporation. In field trials, when microbasins were included, the LEPA system measured application efficiencies averaging greater than 98 percent and distribution efficiency averaging 96 percent with runoff from both irrigation and rainwater essentially eliminated. The LEPA system has been shown also to be superior in water use efficiency and energy saving potential. For the High Plains, the results suggest an economic incentive for conversion to LEPA irrigation.

Drip irrigation is the newest and most efficient of the large-scale irrigation systems. Water is delivered through a network of plastic pipes extended throughout the area to be irrigated. Water is delivered through perforated tubing or emitters on the surface or underground. Low flowrates are used, allowing the water to drip onto or into the soil with little or no ponding on the surface. One difficulty is that low discharge rates require small outlets which are easily plugged, so the equipment needs to be inspected frequently. Drip irrigation systems are usually more costly to install than others, but they offer two important advantages: 1) weed problems are minimal because the surface doesn't get watered, and 2) nutrients and pesticides can be metered in a most efficient way.

Recently developed irrigation technology can increase irrigation efficiency and conserve water. One such development is a device called Hydroturf(TM) produced by the Hydrodyne Corporation and Arco Solar, Inc. This device automatically regulates water delivery to achieve the desired soil moisture level. The system consists of the Hydroturf(TM) controller, a special moisture sensor, an Arco Solar Genesis® thin film module and a nickel cadmium battery. Neither external power source nor control clock are needed. The system can deactivate the watering in mid-cycle if the sensor indicates the desired moisture level has been reached. The sensor is implanted at or near the root zone of the plants and is especially well-suited for subsurface irrigation.

Another high-tech device to enable farmers to save water and energy is the photovoltaic-powered valve activator of a semi-automatic water management system made available by P & R Surge Systems and Arco Solar, Inc. With this equipment the farmer can program the valve controller of an irrigation system so that it automatically optimizes both the frequency and duration of water surges. According to P & R design engineer Robert Bruno, this device can provide water savings of up to 30 percent and energy savings from reduced pumping sufficient to help pay for the system quickly.

An even more sophisticated irrigation scheduling device, called the Scheduler(TM), has been developed and marketed by Standard Oil Engineered Materials, a unit of the original Standard Oil Company. The Scheduler(TM) utilizes infrared radiation, humidity, light intensity, and ambient temperature to automatically calculate a "stress index." This tells whether or not the crop is in water stress well before it is obvious to the naked eye, providing greater sensitivity than methods based on soil moisture alone. Worldwide university research has shown that infrared thermometry readings of plant foliage indicate when a plant is under stress. Farmers know that efficient, precise scheduling of irrigation can result in water savings, improved crop yields, and higher quality crops, but until the Scheduler™ became available it was not possible to easily correlate infrared and atmospheric readings with efficient irrigation timing. In the university studies the plant stress was computer-calculated after separate readings were taken with expensive equipment. The Scheduler™ has a built-in computer that enables it to do the calculations automatically from readings taken by the infrared thermometer and sensors for air temperature, relative humidity, and sunlight intensity. This plant-stress monitor sells for about $4,200 and is being used mainly on the large farms of the High Plains and California.

There are situations with some crops where stressing the plant at the appropriate time can improve the quality of the crop or help it ward off disease; this is true for premium wine grapes. The Scheduler™ has been tested extensively in California and Washington state for this use. Appropriate stressing of a vine by properly scheduling irrigation early in the growing season can reduce or eliminate botrytis rot by causing the plant to form looser clusters. When the cluster is too tight, some of the grapes rupture, providing a breeding ground for the fungus, eventually ruining the entire cluster. Until the Scheduler™ was available, there was no convenient way to know when to stress the vines or by how much.

Irrigation with Saline Waters

The terms saline and brackish generally are used to describe waters which taste salty or unpleasant because of dissolved salts. As used in agriculture the terms are usually quantified as to the amount of dissolved salts present measured in parts per million (ppm). Although saline or brackish water is frequently available, it is rarely used in irrigation because of its adverse effects on plants. This is beginning to change because evidence shows that some crops, under favorable conditions and proper management, can in fact be irrigated with saline water. New salt-resistant varieties have been bred. Also, the need for efficient drainage and avoidance of salt build-up by leaching are better understood. A better understanding of the physiology of salt damage and inhibition of growth is available now. Improved irrigation systems, such as drip irrigation, have been developed, and monitoring equipment for regulating soil moisture and salinity is available. In some locations irrigation waters with up to 3,000 ppm (mg/l) of total solids are being utilized routinely and economically to grow a wide spectrum of crops. Some nonconventional crops have been cultivated successfully using irrigation waters up to 10,000 ppm. There are even plant species which thrive in highly saline waters up to that of sea water, 35,000 ppm. These plants are known as halophytes (salt-loving) of which there are many species. Several halophyte melons and forage crops are grown in limited commercial quantities now, and several other plants are being studied extensively and are nearing commercialization.

Research in this field is still at an early stage, but important studies are being conducted on the agricultural use of saline waters in the Negev Desert of Israel, the Medjedah Valley in Tunisia, in the American Southwest, and in Australia. These projects are highly worthwhile because desalinization of salt water for agricultural use is not cost-effective. However, ample quantities of saline surface and groundwater are available in many regions. Most crops will tolerate irrigation waters whose total dissolved solids are less than 600 ppm. Water containing 500 to 1,500 ppm total solids can be used on all but the most salt-sensitive crops, if leaching and drainage are adequate. Water containing 3,000 to 5,000 ppm total solids can be used successfully only with the so-called salt-tolerant crops such as barley, cotton, wheat, sugarbeets, rye grass, Bermuda grass, asparagus, some melons, and trees such as date palm, pomegranate, pistachio, and olive.

A specific example will illustrate the potential for agricultural utilization of saline water. In the previous chapter we pointed out the problems caused in the Imperial Valley of California by the discharge of large volumes of saline drainage water into the Salton Sea. The average salt concentration of the drainage water is about 3,500 ppm. Reuse of this drainage water could reduce disposal problems and provide more water for irrigation. The same strategy could relieve similar problems in the San Joaquin Valley.

James D. Rhoades of the University of California at Riverside has shown that these brackish drainage waters can be used effectively with properly adapted management practices involving successive irrigation of a sequence of crops of increasing salt tolerance. The strategy has been field tested in both the Imperial and the San Joaquin valleys. In the Imperial Valley two cropping patterns are being studied. One is a two-year successive crop rotation consisting of wheat, sugarbeets, and melons. Colorado River water of 900 ppm total dissolved salts is being used in the preplant and early irrigations of wheat and sugarbeets and for all irrigations of the melons. All of the other irrigations are with drainage water of 3,500 ppm total solids. The other cropping pattern is a block rotation consisting of salt-tolerant cotton for two years, followed by wheat of intermediate salt tolerance and then by more salt-sensitive alfalfa for a block of several years. Drainage water of 3,500 ppm total solids is used for most of the cotton irrigations. Starting with the wheat crop, only Colorado River water (900 ppm) is used. After the wheat crop, desalinization of the soil is sufficient to grow the alfalfa without loss of yield. Preliminary results indicate that no yield loss occurred in any of the crops where drainage water was substituted for the normal irrigation water.

In the San Joaquin Valley field trial, cotton was irrigated with saline water of 6,000 ppm total solids after seedling establishment using California aqueduct water of 300 ppm total dissolved solids. California aqueduct water was used then to grow wheat and to desalinize the soil. Sugarbeets were then grown under the cycle strategy, followed by guar and cotton. The yields were good and compared very favorably with those obtained when irrigating with aqueduct water only.

To date, most farmers have been reluctant to use saline drainage water for irrigation because they think it restricts them to growing salt-tolerant crops and the use of special management practices. This is true, of course, but it seems likely that as these advanced management strategies become better known and the problems of drainwater disposal mount, more farmers will take advantage of this option. Meanwhile, research in the use of brackish and saline waters continues with enthusiasm because of the success achieved to date, and because the potential for additional reward is so great. Many arid lands are underlain with saline water aquifers. The Israelis are using these waters, many ranging from 3,000 to 5,000 ppm, to grow crops by utilizing drip irrigation. Normally the use of such salty water would quickly salinize the soil to such an extent as to make it unfit for crop production. However, by using drip irrigation, applying the water directly to the root zone, the accumulated salts collect outside the root zone and the plant never experiences a salinity higher than that of the water used in the irrigation.

The ultimate achievement will be to use seawater (about 35,000 ppm total solids) in agriculture. While economically feasible in the long term, seawater irrigation is still a distant prospect, with encouraging initial results having been reported already by workers at the University of California at Davis and at the University of Arizona. They have selected salt-resistant strains of barley and several halophytes, and are growing them with seawater irrigation. Researchers at Ben-Gurion University of the Negev in Israel have grown desert xerophytes with ocean water in fields at the sea coast. The Environmental Research Laboratory of the University of Arizona has several halophytes under extensive study on the sea coasts of Mexico and the United Arab Emirates. In addition, several private companies are investigating halophytes. Clearly, much more needs to be done to enhance the use of saline water for irrigation, for the approach is most promising. Some particularly exciting halophytes are described in chapter 4.

Improved Land Preparation Methods

Tillage practices exert a tremendous influence on air, water, and decay in soil. Air is necessary for all beneficial soil life, root proliferation, and the uptake of nutrients and water. The amount of water, too little or too much, can effect adversely nutrient availability and soil life. High priority must be given to tillage practices because they can either favorably or adversely affect nutrient release and utilization by the plant. In arid and semiarid lands what rainfall is available is often of high intensity and short duration, so tillage methods are needed to minimize torrential runoff with erosion and maximize infiltration. Many methods have been employed to accomplish this, with variations depending on such factors as topography, soil type, cropping patterns, and the resources required to implement the practice.

One method which has been employed to a limited extent on the American High Plains and in several African countries is known as *basin tillage.* In this practice mounds of soil are placed mechanically at intervals across the furrow to form small basins. Rainfall or irrigation water is held in the basins to allow adequate time for infiltration and to avoid rapid runoff and erosion. The practice has been reported to provide considerable yield advantage as well as soil conservation benefits, but satisfactory machinery for modern, large-scale, trouble-free, efficient farming does not appear to be widely available.

Minimum tillage or *no-till* practices involve leaving crop residues on the surface and leaving the surface rough to increase water infiltration, reduce evaporation, and prevent wind and water erosion. Labor, machinery, and energy requirements are reduced by limiting the tillage operations. Under optimum conditions crop yields may be improved, energy costs may be minimized, and the conservation advantages significant, but successful management of a no-till operation is considerably more complex than farming under more conventional tillage systems. Several factors should be considered: 1) water retention and residue cover can impede soil warm-up in the spring; 2) medium- to heavy-textured soils tend to become increasingly compacted with no-till; 3) residue decay on the surface can result in lowering of subsurface humus; and 4) broadcasting or dribbling fertilizers onto the surface of medium- to heavy-textured soils fails to deliver nutrients deep enough to be effective. According to Donald L. Schrieffer, author of the book *From the Soil Up,* the fertility problem can be solved by introducing some fertilizer along with the seed, and by placing additional nutrients in a band five inches on each side of the row and five inches deep. In addition to supplying nutrients, these five-inch slots serve as aeration chimneys for diffusion of oxygen and carbon dioxide, and they increase water absorption to the roots. Considering these limitations, no-till offers the greatest potential in the southerly latitudes, in areas with well-drained sands, gravels, or high organic soils, and in locations where the topography is conducive to erosion if conventional tillage practices are used. In some instances, herbicides may be required to control weed growth, although researchers at Michigan State University have reported that leaving residues of rye, sorghum, wheat, or barley on a field can keep it 95 percent weed-free for up to a month. On the positive side, the adoption of minimum tillage methods has been credited with a substantial reduction in phosphorus runoff into the lower Great Lakes.

A proprietary modification of minimum tillage practice called *vertical slot mulching* is being studied and promoted by Agrecology, Inc., a research corporation in Kansas City, Missouri. Vertical slot mulching involves shredding the residue of the previous crop and stuffing it into a vertical slot in the soil. Concurrently, a microwatershed is formed so that rainfall or irrigation water is forced into the slot. This allows for deeper storage of moisture, reduced evaporation loss, and reduced runoff. By draining surplus water deeper into the soil, aeration is improved, especially when rains are excessive. During the early growing season, the improved drainage means a warmer seedbed which accelerates germination and early growth. The watershed crown provides solid footing for tractor and planter to offer better field access during wet seasons. With the farm equipment traveling on the preselected set of watershed crowns, compaction is confined and reduced to a minimum. The farm machinery required for Agrecology, Inc. practices is not widely available, but may be in the near future. Vertical slot mulching is but one of the innovations proposed by this research group to revolutionize agriculture, food, and energy production in the United States and abroad.

Intercropping

Returning crop residues to the soil is beneficial and worthwhile, but at best it restores only a part of what has been removed from the soil. Leguminous plants and trees with their nitrogen-fixing bacteria gather free nitrogen from the air and fix it in the soil in a form plants can utilize. Intercropping is an agricultural practice in which two or more crops, usually including a legume, are intermingled in the same field. This practice has become popular among organic gardeners in the American West, but it has been common practice for centuries by African farmers. These farmers know, for example, that when corn is grown in the same field with cowpeas (also known as black-eyed peas), the yield will be more than double what it would be if corn followed corn with no added fertilizer.

Most of West Africa's farmland is intercropped. For many years, Western advisors and donor agencies considered intercropping to be a primitive practice which should have been replaced with neat monocropping. The latter made it easier to apply fertilizer and pesticides, and simplified mechanical harvesting. However, results of research studies in recent years support the peasants and show that in the African environment intercropping is superior to monocropping. Most African smallholders cannot afford expensive inputs such as pesticides and chemical fertilizers, imported machinery or the energy required to run it. Intercropping and mulching provide a beneficial protective cover of vegetation, increase water infiltration, prevent erosion, save labor on weeding, and reduce problems with pests and diseases.

Intercropping often combines crops of different heights, maturity periods, root depths, etc., to minimize competition between plants for the available light, water, and nutrients. With intercropping the total output for a given area is usually greater than with sole cropping. In most cases crops have different maturity times, so the availability of food from the harvest is spread more evenly. And in Africa's unpredictable environment, intercropping can provide insurance against total crop failure, because while climatic conditions may be disastrous for one crop, the other may survive or even prosper. African farmers are probably the world's experts on intercropping, but scientific research is needed to build upon and improve the practice. Different combinations of planting times, maturity periods, harvesting times, densities, and arrangements of plants all need to be systematically studied. New breeds and techniques need to be developed specifically for intercropping. For example, a low-spreading cowpea intercropped with a tall type of corn will yield much more than a bushy cowpea with a short, stocky strain of corn.

Research needs to be directed toward the breeding of improved strains of legumes and their associated rhizobia. Patterns of intercropping and rotations need to be developed to make maximum use of nitrogen fixation. The judicious use of chemical fertilizers should also be investigated. Studies in Niger show that modest application of phosphate produces spectacular increases in yield. For example, application of only about fourteen pounds of triple superphosphate per acre doubles millet yields and only eight to ten pounds per acre triples the yield of cowpeas. Most African soils are deficient in phosphorus, the element required for root development. Deeper roots can reach deeper moisture, thus offering protection against drought and enabling the plant to utilize nutrients located deeper in the soil.

Alley Cropping

In Africa and other parts of the Third World deforestation is a major cause of land deterioration and the loss of fuelwood, the main source of energy for these regions. In many areas where forests are depleted or gone completely, animal dung substitutes as the fuel for cooking and heating. Consequently, an important natural fertilizer is lost as a factor in maintaining soil productivity. An agricultural practice known as *alley cropping* appears to be the most promising approach to agroforestry in Africa. Food crops are grown between hedgerows of fast-growing leguminous trees. In addition to the nitrogen fixation provided by the trees, the hedgerows are pruned and the leaves used as mulch or fodder. The stems provide fuelwood and stakes.

The choice of trees for alley cropping is important. They need to have deep roots to avoid competition with food crops for water and nutrients. They should be fast-growing and vigorous nitrogen-fixers, so they can provide protein-rich fodder for livestock and nitrogen-rich organic matter for the soil. One of the best is *Leucaena leucocephala*, a leguminous tree from southern Mexico. Alley cropping is superior to fallowing, allowing the same field to be cultivated continuously without the need for chemical fertilizers and improves rather than degrades the soil. This promising agricultural practice was developed at the International Institute of Tropical Agriculture in Ibadan, Nigeria.

Sustainable Agriculture

High-yielding agriculture, based on annual monocultures, tends to reduce the long-term potential for land to produce food. The Land Institute, located near Salina, Kansas, is a private, nonprofit education-research organization, founded and directed by Dr. Wes Jackson. Realizing that the American landscape is losing soil to erosion at an alarming rate, Jackson emphasizes a sustainable agriculture based on the original American prairie as a model. This is an agricultural system less dependent upon fossil fuels and chemicals and more conserving of water and soil. The researchers at the Land Institute design and conduct experiments which they hope will lead eventually to a sustainable agriculture based on high seed-yielding, herbaceous perennial mixtures. Their current research is directed toward answering four basic questions: 1) Can perennialism and high seed yield go together? 2) Can polyculture of perennial plants outyield the same species in monoculture? 3) Can a polyculture of perennials maintain itself through nitrogen fixation and solar energy? 4) Can such an ecosystem control weeds and avoid epidemics of insects and pathogens? This is difficult, long-range research requiring a unique application of ecological principles to agriculture.

Experimental Techniques for Water Conservation

Water is lost by percolating away too rapidly for plant growth, by rapid evaporation, and by transpiration. Methods for reducing water loss by these mechanisms are being studied. Sandy soils are common in many arid regions and the scant rainfall quickly percolates to depths out of reach for most crops. If irrigation water is not available at all or only available in insufficient quantity, such land is of low productivity. Techniques are being developed now to provide artificial underground moisture barriers to keep water and nutrients from percolating below the root zone of desirable crops. The underground moisture barriers can be made of asphalt, plastic, or any durable water-impervious material placed about two feet below the soil surface with gaps every 500 feet or so for drainage. The barriers can be installed by removing the topsoil, hand-placing the moisture barrier, then refilling the area. However, machines are available that can install barriers without excavation. Sand can become a highly productive soil when its low capacity for retaining water is corrected. Field experiments have been conducted using these moisture barriers with field crops and vegetables in Egypt, South Africa, East Africa, the United States, and Taiwan. The cost of installing an underground moisture barrier is relatively high, so it is feasible to use only where crop production on marginal land is desperately needed or where high-value crops can be grown.

A significant portion of rainfall or irrigation water is lost by evaporation. Studies are being conducted with *hydrophilic soil amendments* to minimize evaporation losses. Soils mixed with these hydrophilic (water-attracting) chemicals become sponge-like, trapping water and making it available to plant roots. This can be especially important for increasing available water in sandy soils. Plant roots grow into and around the water-swollen hydrophilic material and extract water and nutrients from it as required. The water-holding capacity of sandy soil can be doubled by incorporating about 5 percent of crushed brown coal into the surface layer. This not only increases available moisture, but promotes uniform soil temperature and stimulates earlier crop maturity.

Experimental hydrophilic chemicals have been developed which can absorb many times their own weight of water. Hydrophilic starch copolymers, called "super slurpers," that absorb up to 1,500 times their own weight of water are being studied at the Northern Regional Research Laboratory in Peoria, Illinois. These are of dubious value in agriculture because research indicates that polymers capable of absorbing over 60 times their weight in water compete with the plant for available water. An Australian firm,

Chemical Discoveries, Ltd., produces two promising soil amendments. One, called Agrosoke™, absorbs 30–40 times its weight in water, holding it tightly enough to prevent evaporation, but releasing it to plant root follicles. Agrosoke™ is a nondegradable synthetic anionic copolymer of acrylamide and is marketed as free-flowing white granules which become sticky when exposed to water. This property enables the polymer to bind soil in such a way as to stabilize it to erosion while increasing aeration porosity and minimizing compaction. Fertilizers and pesticides can be incorporated for slow and efficient release. The second product, called Erosel™, was developed specifically to provide protection of soil from water and wind erosion. This polymer, once incorporated with soil, forms a crust which prevents erosion, but allows absorption of water and oxygen, prevents evaporation of water, and encourages rapid germination when cover crops are sown at the same time Erosel™ is mixed into the soil. Seedlings are protected during growth and their root systems offer additional protection against erosion. This material has many potential uses, including protection of river beds and sea shores, road shoulders, irrigation ditches, and drainage channels. Application is simple. The free-flowing micro-granules are spread on the surface and raked gently, then sprayed with water for about five minutes. Only about two ounces per square yard is sufficient for most uses. More can be applied if a harder surface is required, as in the case of irrigation channels. In the United States another polyacrylamide soil conditioner called Water Grabber™ is marketed by FP Products, Inc. of Atlanta, Georgia. The role of these compounds in arid-land agriculture is still experimental but appears very promising.

Very little of the water absorbed by the roots of a plant is incorporated into the plant cells. Nearly all of the water that moves up through a plant passes into the atmosphere as water vapor. This process is called transpiration. One acre of growing vegetation can transpire as much as 10,000 gallons of water per day. If a way could be found to reduce transpiration losses without harming the plant, substantial reductions in water requirement would result. Research devoted to this objective has yielded some encouraging results. Transpiration losses can be reduced by destroying unwanted phreatrophytes (plants that transpire very efficiently), by breeding varieties that transpire less, by reducing air movement over a crop, or by removing unproductive leaves either physically or with defoliants. But the most interesting approach is through the use of *chemical antitranspirants.*

Water is transpired out of plants through stomata—pores on the leaf surface through which oxygen, carbon dioxide, and water vapor pass. Chemical antitranspirants can be sprayed onto the stomata-bearing leaf surfaces to reduce water passage. Preliminary findings indicate that currently available antitransprirants can reduce water losses by 40 percent in some cases. One example of this approach is presented in figure 3.2. The most promising of the chemical antitranspirants are certain alkenylsuccinic acids and abscisic acid. These compounds work by closing stomata, by forming a film over the stomata, or by cooling the leaf with a reflecting coat which reduces the amount of solar energy absorbed. Reducing transpiration lowers a crop's need for water, but there are limitations to this approach because the stomata are necessary for the intake of carbon dioxide required for photosynthesis and growth. The available antitranspirants do interfere with carbon dioxide movement, so they are most useful where water conservation is more important than maximum plant growth.

Genetic Engineering

This term is used to mean alteration of the genetic components of organisms whether it is done by selective breeding according to the fundamental laws of inheritance as deduced by Mendel in 1866 or by recombinant DNA technology made possible by the discovery of the double helix structure of deoxyribonucleic acid (DNA) by Watson and Crick in 1953. Major contributions have been made, and are still being made, to agriculture by traditional plant and animal breeding techniques. These techniques involve the introduction of desired genetic traits and elimination of undesired traits through natural reproduction and selection of the desired offspring. Recombinant DNA technology (gene-splicing) offers the same potential for improving plants and animals as does conventional breeding techniques, but ideally can develop new strains more rapidly. For example, it takes up to twelve generations or about six years to produce a new insect-resistant strain of tomatoes by conventional breeding, whereas gene-splicing can accomplish the same result in four generations or two years. In addition, the new biotechnology offers the capability for introducing desired

Fig. 3.2. Matched sugarbeet plants, one treated with antitranspirant, were irrigated to saturation and then allowed to grow without further water. Three days later the treated plant still retains its moisture and turgor; the other has wilted. Courtesy of National Academy of Sciences. Photo by R. M. Hagan.

genes from exotic sources, a feat which is not possible in conventional breeding.

The two techniques are having a profound effect on agriculture. For developed countries that produce more food than they need, the major impact is in increasing efficiency of production and the production of new substances which could not be made or were too costly to make by traditional means. For Third World countries that need to produce more food, genetic engineering or a combination of genetic engineering and conventional breeding may provide the needed increase.

Livestock production in arid and semiarid lands is important because vast areas cannot or should not be utilized for any other agricultural use than grazing. These areas support about one-half of the world's cattle, one-third of the sheep, two-thirds of the goats, and virtually all of the camels. Most of these areas are faced with major problems of overgrazing, low reproductive rates, high susceptibility to disease, and inferior quality of meat and milk. There is a serious need to develop more productive breeds that are better capable of utilizing the semiarid pastures more efficiently. The pastures need to be improved to provide better forage. Conventional methods for improving the quality of the herds would require many generations of cross-breeding and selection to obtain improved characteristics, but employment of the biotechnology technique of embryo transfer offers the possibility of accomplishing the desired results in a much shorter period of time. Exceptional cows selected for high productivity, disease resistance, or any other desirable trait can be treated with hormones to induce superovulation. Artificial insemination with sperm from prize bulls will yield seven to ten embryos which can be removed and implanted in ordinary cows serving as surrogate mothers. Twenty to thirty calves could be obtained each year, each having the genetic characteristics of the superior cow. Making use of this approach could induce rapid improvement of livestock to provide high-quality protein to populations in arid and semiarid regions.

Other applications of genetic engineering in animals have led to useful diagnostic tests and treatments for certain diseases. For example, a genetically engineered pseudo-rabies vaccine to control deadly outbreaks of this disease is already on the market. Development of recombinant DNA vaccines for Rift Valley fever virus, trichinosis, and mastitis is in progress. In plants, the new technology has provided a means of reducing frost damage, protection against insect attacks, resistance to herbicides, and production of hybrids which could not be produced by sexual reproduction. Tomato plants altered by genetic engineering to resist damage from tobacco mosaic virus, glyphosate herbicide (Roundup™) and certain harmful insects are being tested in Illinois. In California and Florida the killed cells of *Pseudomonas* bacteria containing genes from *Bacillus thuringiensis* (B.t.) have been used as a bacterial insecticide on lettuce. Countless additional possibilities exist for improving both plants and animals using this new methodology.

While genetic engineering offers great advances for agriculture, there are environmental risks associated with it. The principal concerns are 1) that a genetically engineered microorganism might be so aggressive as to multiply out of control and displace naturally occurring competitive organisms, and 2) that an engineered plant or animal might have an unsuspected defect that would not be detected until after the species had been widely grown. These risks exist also for organisms modified by conventional breeding and by naturally occurring mutations. To date, no national legislative control has been enacted, but regulation by the Environmental Protection Agency, the Food and Drug Administration, and the U.S. Department of Agriculture is underway based on regulatory authority derived from laws enacted for other purposes.

Genetic engineering research is expensive and is hampered by deficiencies in fundamental knowledge. To deal with these problems there is a tendency for universities to supplement their research budgets in this field by entering into agreements with private industry to support joint programs. In spite of some objections and the controversy inherent in any such arrangement, the U.S. Patent Office is granting patents on genetically engineered processes and products. This should stimulate research by providing a means for recovery of the high costs of research and development.

SELECTED INFORMATION SOURCES

Abelson, P. H. 1983. "Biotechnology: An Overview." *Science* 219:611–613.

Arco Solar News. 1987. *Solar Electricity—Companion to Water Conservation,* Spring/Summer:16.

Barton, K. A., and W. J. Brill. 1983. "Prospects in Plant Genetic Engineering." *Science* 219:671–676.

Council for Agricultural Science and Technology. 1982. *Water Use in Agriculture: Now and for the Future,* Report no. 95.

Dregne, H. E., W. O. Willis, and M. K. Adams. 1988. "The Metamorphosis of the 'The Great American Desert.'" *Science of Food and Agriculture* 6:2–7.

Ellis, J. R., R. D. Lacewell, and D. R. Reneau. 1985. "Economic Implications of Water-Related Technologies for Agriculture, Texas High Plains." Texas Agricultural Experiment Station, Texas A. and M. University System, College Station, Texas, MP-1577, July.

Glaaser, K. L., P. J. Wright, and G. P. Leversha. 1987. "Landleveling Using Lasers." *Agricultural International* 39:98–100.
Haggin, J. 1988. "Monsanto Uses Genetic Engineering to Solve Agricultural Problems." *Chemical and Engineering News* 66(February 15):28–36.
Harrison, P. 1987. *The Greening of Africa—Breaking Through in the Battle for Land and Food, An International Institute for Environment and Development—Earthscan Study.* Glasgow: Penguin Books.
Kay, M. G., W. Stevens, and M. K. V. Carr. 1985. "The Prospects for Small Scale Irrigation in Sub-Saharan Africa." *Outlook on Agriculture* 14:115–121.
Lacewell, R. D., and G. S. Collins. 1982. "Implications and Management Alternatives for Western Irrigated Agriculture." Paper presented at the Conference on Impacts of Limited Water for Agriculture in the Arid West, Asilomar, California, September 20–October 1.
Lyle, W. M., and J. P. Bordovsky. 1986. "Chemical Application with the Multifunction LEPA System." *Transactions of the American Society of Agricultural Engineers* 29:1699–1706.
Lyle, W. M., and J. P. Bordovsky. 1986. "Multifunction Irrigation System Development." *Transactions of the American Society of Agricultural Engineers* 29:512–516.
Moss, V. 1987. "Tracking Biotech's Progress." *Chemical Business* December, 23–25.
National Academy of Sciences. 1974. *More Water for Arid Lands: Promising Technologies and Research Opportunities.* Washington, D.C.: Commission on International Relations.
Pimentel, D. 1987. "Down on the Farm: Genetic Engineering Meets Ecology." *Technology Review* 90:24–30.
Rhoades, J. D. 1984. "Use of Saline Water for Irrigation." *California Agriculture* 38:42–43.
Riotte, L. 1984. *Carrots Love Tomatoes—Secrets of Companion Planting for Successful Gardening.* Pownal, Vt.: Garden Way Publishing.
Schechter, J. 1985. "New Frontiers in Desert Research," In Y. Gradus, ed., *Desert Development, Man and Technology in Sparselands,* pp. 287–309. Dordrecht: D. Reidel.
Schrieffer, D. L. 1984. *From the Soil Up.* Des Moines, Ia.: Wallace-Homestead.
Seltzer, R. 1988. "Engineered Organisms: Small-Scale Field Tests Pose Little Risk." *Chemical and Engineering News,* May 9, p. 4.
Steiner, K. G. 1984. *Intercropping in Tropical Smallholder Agriculture with Special Reference to West Africa.* Deutsche Gesellschaft fur Technische Zusammenarbeit.
Stinson, S. 1988. "Biotech Outlook: Congressmen Push for More Support." *Chemical and Engineering News* 66:4–5.

FOUR

Food Crops for Arid Lands

> The greatest service which can be rendered any country is to add a useful plant to its culture. —Thomas Jefferson

According to Noel Vietmeyer, a professional associate of the National Research Council in Washington, D.C., more than 20,000 edible plants are known. Throughout human history some 3,000 species may have been used by humanity as food, but today the world's food supply is dominated by only 20 or so species. The world's annual production of the top 30 plants is shown in figure 4.1. Five crops—sugar, wheat, rice, corn, and potato—provide more tonnage than all the rest put together. It is noteworthy that not one of these plants will grow well in arid land without irrigation. To meet the challenge of rapidly growing population with rapid deterioration of arable land by desertification, salinization, and deforestation, we need to make marginal lands more productive. We need to expand our agricultural resource base to relieve our precarious dependence on so few crop plants. This problem has been investigated by the National Research Council and other organizations. Considerable enthusiasm has been generated for genetic engineering to solve the problem with the production of "humanmade organisms." While this approach offers promise and should be encouraged, the huge number of lesser-known food crops in nature should not be neglected. There are many obscure

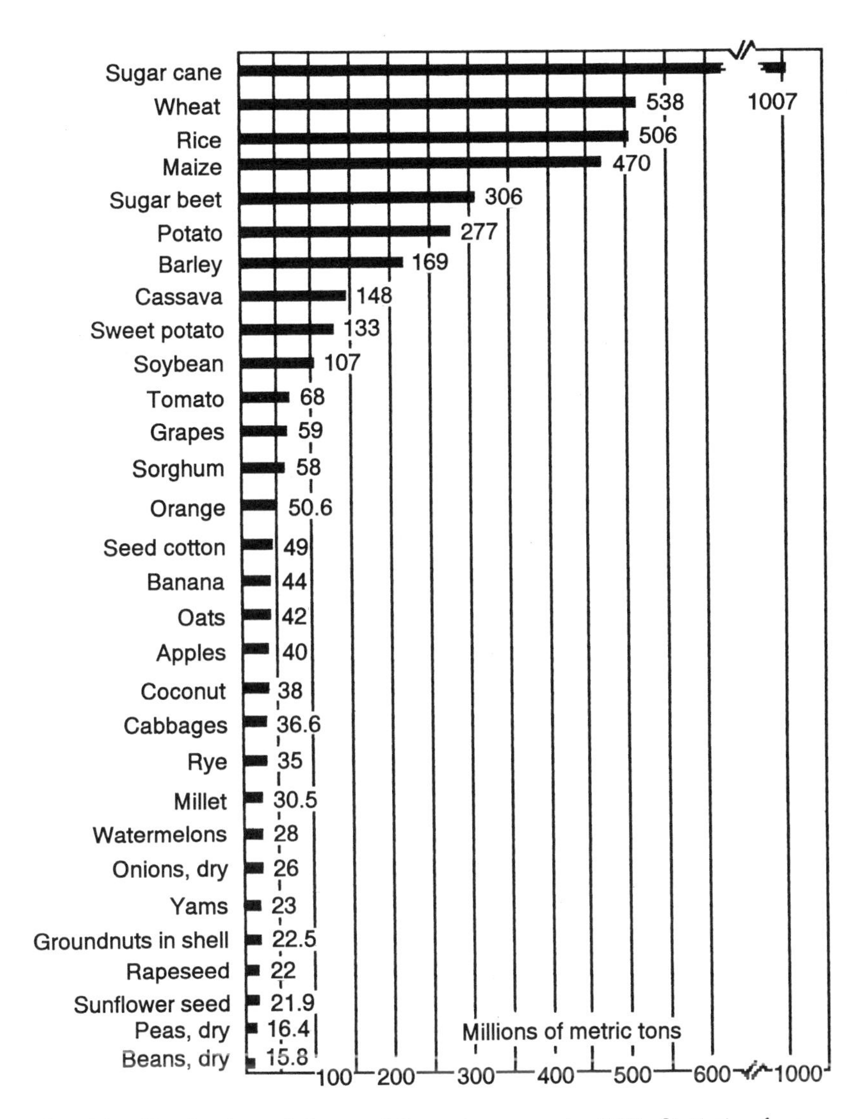

Fig. 4.1. Production of the world's major crops in 1989. Statistics from the Food and Agriculture Organization of the United Nations (FAO Production Yearbook), vol. 43, 1989, as obtained from UNIPUB, the exclusive U.S. distributor for FAO publications. Bar graph prepared by Justin Bond Breese.

plants which are naturally adapted to arid regions, and they offer great potential for feeding indigenous populations in arid lands while preserving the soil and restoring barren lands to productivity. The purpose of this chapter is to list and describe a number of these plants which deserve consideration for agronomic research and development.

Grain Amaranth

Amaranth is the generic term for a group of annual herbs which are vigorous growers and widely distributed. Some members of the group are grown only as ornamental plants, but others serve as food crops in various parts of the world. In the hot, humid regions of Africa, Malaysia, Indonesia, southern China, southern India, and in the Caribbean islands vegetable amaranth species are grown and prized as boiled salad greens. The cooked leaves are flavorful and nutritious. For arid regions the grain amaranths are of greater interest.

Prior to the Spanish conquests of the New World, grain amaranth was one of the basic food crops of the Aztec and Incan cultures, ranking with corn and beans in importance. Many thousands of acres in the Andes, in Mexico, and in what is now Arizona were used to raise the tall, reddish plants which yielded tons of the tiny amaranth seeds. When heated, they burst or pop to take on a flavor similar to that of popcorn. The Aztecs used popped amaranth in pagan ceremonies including human sacrifice. This shocked the conquering Spanish conquistadors and they banned cultivation of amaranth, forcing it into obscurity. Fortunately, a few farmers in isolated regions of Mexico and South America preserved the ancient tradition of amaranth culture. In the 1970s W. J. Downton, an Australian researcher, obtained a small amount of amaranth grain and learned that the tiny seeds contained unusually high levels of both total protein and the essential amino acid lysine. Most plant proteins, including those of corn, wheat, and rice, are deficient in this nutritionally essential amino acid. There is now growing interest among cereal researchers and agronomists in this ancient crop so that it may well be making a serious comeback.

Grain amaranths represent a promising crop for hot and dry regions because they utilize an especially efficient type of photosynthesis to convert nutrients from the soil and air, water, and energy from the sun into plant tissues. Technically this process is known as the C4 carbon-fixation pathway, which is a modification of the usual photosynthetic process. The relatively few plants which use the C4 pathway can convert a higher ratio of atmospheric carbon dioxide to plant sugars per unit of water lost than those utilizing the classical C3 (Calvin cycle) pathway. Plants using the C4 pathway are able to maintain relatively high rates of carbon dioxide fixation even when their stomata are partially closed. When a plant is stressed by drought or salinity, they close their stomata. When stomatal openings are partially closed, water loss by transpiration is reduced. For these reasons, plants such as amaranth perform better than C3 plants under adverse conditions. Their photosynthesis is particularly efficient at high temperatures, in bright sunlight, and under dry conditions. Amaranths can tolerate a considerable lack of water without wilting or dying. For example, research has shown that one species, *Amaranthus caudatus,* achieves peak photosynthetic activity at 104 degrees F. (40 degrees C.). Grain amaranths have been grown in dry-land farming areas receiving as little as 8 inches (about 200 mm.) of rainfall per year.

There are three species of grain amaranths which produce large seedheads containing thousands of tiny edible seeds. *Amaranthus hypochondriacus* and *Amaranthus cruentus* originated in Mexico, Guatemala, and the southwestern United States. *Amaranthus caudatus* is a native of Peru, Ecuador, Bolivia, and northern Argentina. Farmers in isolated regions of Mexico, Central America, and South America still cultivate all three species on a small scale. The amaranths are broad-leafed plants, one of the few non-grass plants that produce edible grain. They are attractive, colorful plants with brilliantly colored leaves, stems, flowers, and seed pods. The seed can be used as a breakfast cereal, it can be popped, parched, or cooked into a gruel. The seeds can be ground into a light-colored flour suitable for breads and other baked goods. However, amaranth grain contains little functional gluten, so it must be blended with wheat flour before it can be used to make yeast-leavened bread "rise." The popped seeds can be mixed with honey to form a tasty confection.

While all grain amaranths originated in the Americas, some time after Columbus *Amaranthus hypochondriacus* made its way through Europe and by the early nineteenth century it had been taken to Africa and Asia. Today it is planted as a grain crop in the mountains of Ethiopia and by hill tribes in widely scattered parts of India, Pakistan, Nepal, Tibet, and China. In a few local areas in the Himalayas this amaranth species is an important food crop where it is eaten as a flat bread. In northern India, just as the Aztec and Maya did centuries ago, the popped grain is mixed with honey to make a confection they call *laddoos.*

After the Australian plant physiologist John Downton reported in 1972 that amaranth seed contains protein of superior quality with a high content of the amino acid lysine, the species attracted increased research attention. Agronomists throughout the world undertook studies to learn how this crop could be commercially produced. Research to determine the influence of climate, soil, pests, and diseases was carried out. The Rodale Research Center in Kutztown, Pennsylvania is leading in the development of amaranth in the United States. This group has been remarkably successful in developing uniform amaranth varieties that can be mechanically sown, cultivated, and harvested under modern farm conditions. Since 1976 more than 1,000 amaranths from all parts of the world have been collected and evaluated. Collaboration with scientists in Africa, Asia, and Latin America has resulted in the development of strains with more uniform and superior growing properties, and with improved baking, milling, popping, and taste qualities. Several commercial seed sources in the United States offer supplies of *Amaranthus cruentus* and *Amaranthus hypochondriacus* seeds for sale. Figures 4.2, 4.3, and 4.4 show examples of amaranth species grown from commercially available seed. More than twenty U.S. farmers are growing grain amaranth using modern farm methods, and several companies are testing the grain in their products. Several small companies offer amaranth products including grain, flour, popped seeds, crackers, pasta, cookies, and granola on the retail market.

A great amount of practical information on growing, harvesting, storage, marketing, sources for amaranth products, a bibliography of popular articles and other useful information is available in the *Amaranth Grain Production Guide* published by the Rodale Press. Beginning in March 1983 the *Amaranth News Letter* has been published quarterly by the editorial office of Archivos Latinoamericanos de Nutricion with the sponsorship of the National Academy of Sciences in Washington, D.C. As stated by Ricardo Bressani, editor-in-chief, in the newsletter's first issue, "This newsletter was precisely conceived to . . . create a network of amaranth

Fig. 4.2. Grain amaranth can be grown in many regions under a wide range of conditions. This plant grown by one of the authors, Dr. Jack W. Hinman.

researchers guided by the common purpose: To speed up the dissemination of knowledge on agricultural, technological and nutritional research on amaranth; to strengthen the research net, finally, to call attention to the potential that these food products offer for human nutrition."* With all of this interest and activity—and positive results—grain amaranth may be on the threshold of at least limited commercial production in the United States.

*Those interested in obtaining this publication may do so by writing to Dr. Ricardo Bressani, Archivos Latinoamericanos de Nutricion, INCAP, Apartado postal 1188, Guatemala, C.A.

Fig. 4.3. Grain amaranth seed head (*Amaranthus* spp.).

Quinoa

Quinoa (*Chenopodium quinoa* Willd) is another broad-leafed annual herb which produces a highly nutritious grain, rich in protein, and like that of amaranth, contains an amino acid composition very favorable for human nutrition because of its high levels of lysine and methionine. Originating in the high Andes country of Peru and Bolivia, quinoa became a staple food of the ancients there, including the Incas. Quinoa is still cultivated in the Andes by peasants, but also by modern farmers. In Peru and Bolivia there are over 100,000 acres devoted to quinoa growing by the Quechua Indians. Cultivation in the United States, however, did not begin until the early 1980s.

Fig. 4.4. A cultivated field of grain amaranth being grown in California. Photo courtesy of Dr. Jess Martineau.

The quinoa plant is an annual herb which grows to a height of three to six feet (one to two meters); it has a shallow root system, a hollow stem, lobed, triangular leaves, and bears numerous small, green, densely clustered flowers. At maturity, quinoa plants have a sorghum-like head loaded with seeds about the size of millet and shaped like a miniature aspirin tablet. Seed color, caused by a saponin coating, varies from black, red, orange, and yellow to white. For human consumption, this bitter-tasting coating must be removed by washing with water. After removal of the saponin coating, the seeds are generally pale yellow to white.

There appears to be no "pure" strain of quinoa. It has been grown for centuries in many locations and under a variety of ecological conditions. In Peru and Bolivia there are quinoa seed banks with over 1,800 ecotype samples of quinoa. Recently Bolivian breeders have selected a relatively saponin-free variety called sajama. This strain requires no washing and is being field tested in Bolivia and Peru. Some ecologists are concerned that the bitter-tasting saponin may protect the seeds from insect and bird damage. They fear that the saponin-free strain may require more protection by chemical pesticides, yet to date there are no pesticides or herbicides approved for use on quinoa.

The edible seed of quinoa has been called one of nature's most perfect foods. It is not only rich in high-quality protein compared to other grains, but also contains more oil, calcium, phosphorus, and iron. In addition, it is a good source of vitamin E and several B vitamins. Quinoa seeds can be prepared whole, like rice, used in soups, and ground into flour for bread and cakes. A breakfast cereal is manufactured in Peru, where it is also fermented into a beer called "chicha." The leaves are used as a green vegetable and for animal fodder. The stalks can be burned for fuel and the saponin-filled washwater can serve as a shampoo, a textile rinse, and as a wetting agent.

Quinoa thrives under conditions where most food crops fail: low rainfall, high altitudes, thin cold air, intense radiation levels, hot sun, freezing temperatures, poor, sandy, and alkaline soils. In 1983 Bolivia was stricken with a severe drought. Crop losses ranged up to 66 percent for potatoes and 54 percent for barley, but quinoa yields were barely affected. At soil temperatures of 45 to 50 degrees F. quinoa germinates within 24 hours and emerges in 3 to 5 days, depending on soil type and moisture content. Optimum precipitation ranges from 30 to 35 inches. When grown in an area which receives less than 20 inches, supplemental irrigation is recommended. Grain yields decrease when conditions are less than optimal, but much less so than in the case of conventional crops such as corn and barley. Quinoa is quite tolerant of light frost (30–32 degrees F.), but a killing frost (20–24 degrees F.) when the plants are in bloom will cause a significant loss. Yields are increased with nitrogen fertilization. A fall rain at harvest time can be a problem, because the seed still in the seed heads will germinate. Planting, tilling, and harvesting, while all done by hand in the high Andes country, can be done with conventional farming equipment. At maturity the sorghum-like seed heads can be combined using a sorghum header attachment.

In 1976 a Bolivian introduced Steve Gorad to quinoa. Gorad was so impressed with it that he contacted South American scientists and quinoa growers, and in 1978 brought 50 pounds of quinoa seeds to the U.S. He cooked and served this to all his friends, and when all of them liked his quinoa, he and Don McKinley decided to market quinoa in the this country. After dealing with a number of problems, and even with the assistance of the late Dr. David Cusack, founder of Sierra Blanca Associates, they gave up efforts to import large quantities of quinoa and in 1983 started growing their own in the San Luis Valley at 8,000-foot elevation in arid central Colorado. The Quinoa Corporation* was formed and today is marketing quinoa grain, pasta, and flour to health food stores, specialty food stores, supermarkets, and restaurants nationwide.

Duane L. Johnson and Robert L. Croissant of Colorado State University Cooperative Extension have been studying quinoa cultivation in the San Luis Valley and have one variety, D407, available for experimental production. This variety has early maturity, a semidwarf growth habit, yellow compact seed heads with medium small kernels, and has given consistent yields of 1,200 lbs./acre. The Talavaya Center, a non-profit research and educational facility** has been researching quinoa under the direction of Emigdio Ballon. This organization offers for sale six quinoa cultivars, including one for low elevation and a grower's handbook.

It seems likely that quinoa could become a popular food item in the U.S. and elsewhere in the foreseable future. It is a potential "new" crop for farmers in arid and semiarid regions throughout the world.

Triticale

In 1876 a Scottish botanist, Alexander Stephen Wilson, took pollen from rye plants and used it to fertilize wheat flowers to produce the first humanmade grain. The hybrid plants he grew from the resulting seeds were exciting to botanists, but of little interest to farmers because they could not reproduce themselves. The next fundamental advance came in 1938 when the Swedish plant geneticist Arne Muntzing applied the colchicine treatment to his wheat/rye hybrids and found that it transformed them into plants that produced viable seeds. This new plant, neither a wheat (*Triticum*) nor a rye (*Secale*) was named *Triticale* (usually pronounced trit-i-kay-lee) by the Austrian plant breeder, Erich Tschermak-Seysenegg. This humanmade grain, developed during the last hundred years, has the potential to become a major staple worldwide.

*P.O. Box 1039, Torrance, California 90505.
**P.O. Box 707, Santa Cruz Station, Santa Cruz, New Mexico 87507.

Triticale combines many of the best qualities of both its parents. From the rye "father" it can have most of rye's resistance to disease, drought tolerance, hardiness, and ability to grow in marginal soils. And from the wheat "mother" it can have most of wheat's qualities for making pastas, pastries, and some breads. Although the levels of nutrients were extremely variable, compared with wheat, triticale had slightly higher levels of most of the nutritious constituents. Twenty years ago scientists in several countries produced triticale varieties which appeared to have commercial promise. Unfortunately, some promoters exaggerated triticale's potential. The cultivars available at that time often set seed poorly, the grains were shriveled and small, and yields were generally disappointing. Word spread that farmers had been "taken." Funding for triticale research all but dried up and the crop got a bad reputation.

Despite the lack of research funds, a few dedicated scientists in North America and Europe continued triticale development programs and by the 1980s the results of their efforts were substantial. In the Mexican state of Sonora, at Ciudad Obregon, the Centro Internacional de Mejoramiento de Maiz y Trigo (CIMMYT) maintained intensive triticale research. There in 1967 a crucial breakthrough occurred accidentally. A triticale plant was fertilized by pollen blown in from nearby plots of dwarf bread wheats. After a few generations of selection it resulted in a new breeding line named Armadillo which helped resolve many of the agronomic problems plaguing triticale. This new strain was high-yielding, disease-resistant, early maturing, the straw was short and stiff to resist wind damage, and the grain was only slightly shriveled.

Encouraged by the Armadillo breakthrough, the International Development Research Center and the Canadian International Development Agency funded a five-year program at CIMMYT and the University of Manitoba to develop triticale strains for use in Third World countries. Enthusiasm rose again and by the end of the 1970s triticales were being tested in 400 locations in 83 countries. By 1989 CIMMYT estimated that triticales were grown on more than 4 million acres worldwide. In France, the Soviet Union, and Poland more than two-thirds of the total of 2.8 million acres were planted in the winter variety of triticale. In the U.S., both spring and winter varieties are grown, but only on about 60,000 acres, and the government allows farmers to grow and harvest triticales on land bank land.

In Australia, fifteen triticale varieties have been released to farmers since 1979. The crop was rapidly adapted, partly because it performed so well on acid soils, so that by 1985 nearly 400,000 acres of triticale were grown. Since 1983 more triticale has been grown by Australian farmers than cotton and rice. Most of the Australian triticale has been used to feed livestock, some has been exported to Southeast Asia, but its use as a human food has begun and a cookbook with more than 100 recipes has been published. The once maligned crop has made a comeback.

Current production of triticales is mostly in industrialized countries, but interest is rising in Asia, Africa, and Latin America. This is most appropriate because triticale varieties are available for problem soils and weather conditions poorly suited for other food grains. Triticale's outstanding tolerance to acidity, alkalinity, salinity, toxic elements such as aluminum and boron, and its resistance to drought, disease, and pests make it a potential food crop for many troubled areas. Although to date most triticale grain has been used to feed livestock, there is no reason to prevent its becoming a popular human food cereal. It adds flavor to most baked goods, and it has a taste that most people like instantly. Triticale behaves like soft-wheat flour in unleavened breads, and the baking process needs no modification. This makes it most appropriate for many countries of Asia, Africa, and Latin America where the staple food is some form of unleavened flat bread such as enjera, chapatis, concha, or tortillas. For the preparation of leavened bread the early varieties of triticale were no match for bread wheat because they had less gluten than wheat. However, by careful selection, CIMMYT researchers found lines with gluten quality equal to that in bread wheats. Leavened bread made with flour from these varieties of triticale rises to normal levels. Unfortunately, the dough made from even the best lines cannot fully compete with wheat because it is more sticky than wheat dough. This is of no concern in home or small-scale bakeries, but in large modern bakeries with high-speed mixers a small amount of wheat flour must be added before the dough will roll off mixers in the required fashion. Triticale pancake mixes and crackers sold in specialty food stores are gaining popularity because of their savory, nutty flavor and North American consumers pay a premium price for triticale in bread and snacks. Triticale mixes well with non-wheat flours. Breads and muffins made with a blend of quality-protein maize and triticale have a nutty flavor and a chewy texture that make their white-flour counterparts seem bland by comparison.

With regard to nutritional performance, trials on living animals have shown triticale to have a biological value 15–20 percent greater than that of wheat. This superiority is probably due to triticale's higher content of the essential amino acids lysine and threonine. Major minerals such as potassium and phosphorus are marginally higher in triticale. This is true also for micronutrient elements such as manganese, iron, sodium, and zinc, but triticale's vitamin content is about the same as wheat's, with niacin the most limiting vitamin in both grains. Starch and bran contents are comparable to those of wheat. Everything considered, triticale is the second-most nutritious grain available today for widespread commercial use. Only oats have a better nutrient content. Grain amaranth and quinoa, as yet, are available only in limited quantities. Clearly, triticale is a promising addition to the world's cereal grains.

Prickly Pears

Few residents of the temperate zones of North America and Europe would expect members of the cactus family to offer economically promising species as a source of human and animal food. Yet these plants have been utilized and appreciated in many semiarid regions of the world for centuries. There are approximately 150 genera and 1,500 species in the Cactaceae family, all of which were originally native to the New World. Columbus introduced one species, *Opuntia ficus-indica,* to Spain on his first return trip from the West Indies. From Spain this cactus spread to North Africa, Italy, Greece, and other Mediterranean countries by the end of the eighteenth century. By 1965 the prickly pear occupied plantings of 247,000 acres in Sicily, 14,800 acres in Sardinia, and somewhere between 150,000 and 200,000 acres in Tunisia. *Opuntia ficus-indica* has been grown in Brazil as a fodder crop for eighty years. Currently, there are some 740,000 acres in cactus plantations in Brazil. Commercial plantings of various *Opuntia* spp. for fruit, fodder, forage, or vegetable exist in Algeria, Argentina, Chile, Mexico, South Africa, and other semiarid regions of Latin America and Africa.

All members of the cactus family have a special carbon dioxide fixation pathway which provides them with a four- to fivefold greater efficiency in converting water and carbon dioxide into plant tissue and dry matter than even the C4 pathway plants such as corn and amaranth. This pathway, known as crassulacean acid metabolism (CAM), makes cacti so water-use efficient that they are especially suitable for arid and semiarid regions.

There are three major growth forms of cacti: the large columnar type, the shrub to small tree-sized prickly pears and chollas, and the small pincushion cacti. The large columnar forms such as the stately saguaro produce sweet fruits, but they grow too slowly to be of economic importance. Of the small pincushion type, at least one species, *Mammillaria microcarpa* Engelm, has been recommended for commercialization because it produces inch-long spineless fruits which are sweet-tart and might be marketed as a strawberry-like specialty crop. Of all the forms, the prickly pears and chollas offer the most promising germplasm for commercial development. Three of the best candidates are *O. lindheimeri* Engelm, a prickly pear native to South Texas, *O. phaeacantha* Engelm, a Sonoran Desert prickly pear with sweet, juicy fruits, and *O. ficus-indica* (L.) Mill., the large, usually spineless, domesticated prickly pear which Columbus introduced to the Old World. Figures 4.5 and 4.6 show prickly pears growing wild in Arizona.

Prickly pears are grown for fruit, as a vegetable crop, and as fodder and forage for livestock. The fruits of prickly pears are called *tunas* by most Latin peoples, varying in size, color, and shape with species and location. The amount of pulp varies from 30 to 60 percent, with the best tasting varieties containing the largest amount of pulp. Red fruits can often be purchased in supermarkets of southern California, yellow-green fruits in Santiago, Chile, delicious light yellow *tuna* are sometimes even served on commercial airlines in Chile. The Texas *O. lindheimeri* produces two-inch purple fruit, and the cultivated *O. ficus-indica* is valued for its big, juicy red fruit. In addition to the ripe fruit, various cottage industries prepare products from prickly pear juices including a *tuna* honey, a molasses-like product, jelly, *tuna* cheese, a taffy-like confection, a fermented drink, and dried *tuna*.

One published report indicated that a Mexican prickly pear plantation using a plant density of about 800 cacti per acre yielded 3.2 metric tons of top quality fruit within four years of its establishment. There appears to be a good supply of commercially acceptable strains for adapting prickly pear fruit farming to different climates and soil conditions. Considering the favorable qualities of prickly pear fruits and the current lively market for exotic produce, customer acceptance would probably not be a problem if prickly pear fruits were commonly available in the U.S. Further, with the water supply problems, the surplus of many traditional crops, and

Fig. 4.5. Prickly pear cactus (*Opuntia* spp.) bearing fruit.

land values in some semiarid regions, cactus farms might provide a welcome increase in income for some landowners.

The tender young pads of prickly pears, called *nopalitos*, are eaten as a vegetable in Mexico and the southwestern United States. It is customary to serve them as a cooked green vegetable during the Lenten season and as a marinated vegetable during the rest of the year. Production of nopalitos is a serious business in the small valley of Milpa Alta located some 15 miles southeast of Mexico City. Here the entire cropping area of this 8,000-foot altitude valley is devoted to nopalito production with the domesticated prickly pear, *O. ficus-indica*. The plantations use a plant density of about 16,000 plants per acre. The first crop can be harvested in 2 or 3 months with well-established plantings yielding 30 to 36 tons per acre. Pads in excess of those needed for the vegetable market

Fig. 4.6. Gathering fruit of wild prickly pear cactus in anticipation of preparing jelly and candy. Photo by Dr. C. Wiley Hinman.

are traded to dairy operators for use as cattle fodder; the pads are exchanged for manure which is the only fertilizer used. The dairy farmers consider the cactus pads important for good lactation, and they believe the pads impart a superior flavor and quality to their milk and butter, for which their customers are willing to pay a premium price. The local farmers claim that changing from growing corn, which was frequently a marginal crop, to growing cactus made them prosperous. This appears to be a good example of converting marginal land into highly productive land simply by growing an ecologically appropriate crop.

In much of south Texas and northeastern Mexico the growing season exceeds 300 days and the calculated average annual rainfall is 15 to 28 inches. But this area suffers from wide variability in annual rainfall with no accurate predictability for any given year. From the standpoint of maintaining a grass crop on these rangelands, three or four out of seven years may be drought years. This unpredictability makes range management difficult and frequently results in overgrazing and degraded rangeland. In this area the prickly pear is widely used as drought emergency feed for cattle because when the grasses are gone the cacti remain succulent and green. The native prickly pear, *O. lindheimeri,* is a spiny cactus, so to make the pads safe for livestock consumption the ranchers burn off the spines with a propane torch known as a "pear burner." Clearly, one possibility for these drought-prone rangelands would be to grow prickly pears as a fodder crop. Spineless varieties could be grown to avoid the need to burn off the spines or remove them by mechanical means.

Since the water-use efficiency of the cactus photosynthetic pathway is greater than that of grasses and legumes, dry matter productivity of cacti should be high. Reliable information on this point is limited, but dry matter productivity in the range of eight tons per acre would seem reasonable. Studies carried out in Tunisia indicate that although pads are often less than 1 percent nitrogen on a dry-weight basis, they respond well to increased fertility. Eighteen pounds per acre of nitrogen, phosphorus, potassium fertilizer induced a 250 percent increase in cactus production. This suggests that cactus production would be improved by interplanting with legumes. Perhaps this explains why the predominant vegetation of the arid and semiarid areas of the southwestern U.S. and northern Mexico is a mixed cactus/tree legume (mesquite) brush ecosystem.

In northeastern Brazil field trials have been conducted on various intercropping regimens with cacti. In the more arid regions prickly pears were intercropped successfully with mesquite (*Prosopis* spp.). In wetter areas, and especially during the rainy season, *Opuntia* and/or *Nopalea* spp. were intercropped with corn, beans, cotton, peanuts, sorghum, and other crops. In drought-prone areas throughout the world this approach offers promise. Prickly pears are especially amenable to intercropping agroforestry and silvopastoral systems with tree legumes such as *Prosopis* and *Leucaena* and annual crops such as sorghum and various types of beans and peas.

Promising Legumes for Arid Lands

Much attention and enormous resources have been devoted to research on grasses such as wheat, rice, and corn, but among the legumes only soybeans and peanuts have received any attention. Legumes are unique among plants in that they harbor nitrogen-fixing bacteria (rhizobia) in their root nodules. This enables them to take gaseous nitrogen from the air and convert it into water-soluble compounds which are important plant nutrients. One acre of an inoculated legume can take as much nitrogen from the air as would be contained in ten tons of manure. These plants require little or no nitrogen fertilizer, they will grow on barren, nitrogen-deficient land, and their residues leave the soil enriched in nitrogen and organic matter. Their fruits and forage are rich in protein, vitamins, and roughage, so they are attractive crops for developing countries. There are some 18,000 species of legumes in the plant world, and many of them thrive in arid and semiarid climates. In addition to the familiar beans, peas, alfalfa, and clover, there are leguminous roots and tubers, shrubs and trees. Few of these potentially important plants are known outside the regions where they grow wild or are cultivated on only a limited and primitive basis.

The *bambara groundnut* (*Voandzeia subterranea*) is a good example of a little-known legume grown throughout sub-Saharan Africa. It produces an edible seed underground like peanuts, but it will thrive in arid regions where peanuts fail and has high resistance to pests and diseases. The nuts are of good flavor, are a well-balanced food, but contain less oil and protein than peanuts. They have a caloric value equal to that of good-quality grain. Africans are said to prefer the bambara groundnut to peanuts. In spite of its desirable qualities and its importance in Africa, it has received virtually no research attention. A photo of a Tanzanian specimen of this potentially important legume is shown in figure 4.7.

The *marama bean* (*Tylosema esculentum*) is another drought-tolerant perennial legume worthy of note. It is native to the Kalahari Desert in southern Africa, and produces both an edible tuber and seeds which are roasted like almonds or pistachios and have a nutty flavor comparable to cashews. The young tubers are baked, boiled, or roasted and have a sweet, pleasant taste, while older tubers are avoided because they are tough and have a bitter taste. The protein content of the seeds is reported to vary between 30

Fig. 4.7. Bambara groundnut (*Voandzeia subterranea*). Courtesy of Agriculture Research Institute, Ukiriguru, Mwanza, Tanzania.

and 39 percent and have an essential amino acid content slightly better than that of soybeans, comparable to that of the milk protein, casein. Very recently, Bower et al. reported on their comprehensive nutritional evaluation of the marama bean. The seed was analyzed for protein, amino acids, oil, fatty acids, fiber, caloric value, trypsin inhibitor, and mineral content. In contrast to what previous investigators had reported, they found somewhat lower values for the essential sulfur-containing amino acids, but they still considered marama protein quality better than that of most legume crops, such as garden beans and peas. They found the marama bean oil to contain 48.5 percent oleic acid, 19.2 percent linoleic, 2.0 percent linolenic, and smaller amounts of medium-chain saturated fatty acids. Proper cooking was recommended to render the bean more palatable and to reduce the trypsin-inhibitor activity. They concluded that for semiarid regions of the world, marama cultivation could prove more valuable than some of the established crops as a source of complete protein, minerals, carbohydrate, and lipid in the diet of both human beings and animals. Nutritional analysis of a tuber from a five-month-old plant by other investigators indicated 2.1 percent protein, 0.14 percent fat, 4.38 percent carbohydrate, 0.42 percent ash, and 92.1 percent water.

The Kalahari Desert, which is the home of the marama bean, receives an average annual rainfall of slightly less than 10 inches, but is subject to wide variations including prolonged droughts and in some years precipitation of 24 inches or more. Although this plant offers great potential in arid land agriculture, no agronomic research was undertaken until the early 1980s. Very recently, A. M. Powell reported on studies conducted in the Chihuahuan Desert Research Institute and Sul Ross State University of Alpine, Texas, where marama beans were grown successfully in a limited experimental situation. The plants exhibited vigorous perennial growth and yielded a healthy seed crop in about four and a half years. This is an important development because the native African seed sources have been declining. Using an ample supply of seeds produced in the U.S., evaluation of the potential of the marama bean should be completed soon. Seed propagation and transplants of greenhouse seedlings are being studied in rangeland experiments, and use of the marama bean as the legume in intercropping warrants further research. In addition, the market potential for the nuts and the oil needs to be explored.

The *tepary bean* (*Phaseolus acutifolius*) is a desert legume which has been grown for food by the Indians of the southwestern United States and northwestern Mexico for centuries. It is an "ephemeral" annual which thrives in arid and hot climates and survives in poor soil. It matures so quickly that one desert rainstorm can provide enough moisture to enable the tepary bean to set its flowers and bring seeds to maturity. The tepary is smaller than common beans and comes in a variety of colors, although

Fig. 4.8. Tepary bean (*Phaseolus acutifolius*). Drawing courtesy of P. Mirocha of Office of Arid Lands Studies, University of Arizona.

most of the domestic varieties are either white or light brown. Wild teparies are smaller than the domesticated beans and are usually gray in color with brownish mottling. They are cultivated for personal use, mainly by the Indians, and on a small scale commercially for local markets in Arizona and the Mexican state of Sonora. Both manual and mechanized methods of farming are used. Tra-

Fig. 4.9. Tepary bean plant (*Phaseolus acutifolius*). Picture courtesy of Office of Arid Lands Studies, University of Arizona.

ditionally, teparies are planted in late July with late summer rains capable of producing runoff. The plants emerge from the hot, moist soil in four to eight days and within four to six weeks begin to set small white or lilac-colored, self-pollinating blossoms. In less than 45 days they begin to fruit and in less than 70 to 75 days nearly all the pods are ripe. Under similar conditions, most of the pods of pinto beans would still be green. Illustrations of the tepary bean plant are provided in figures 4.8 and 4.9.

In some areas, especially where irrigation water is available, teparies are sown in March or April and a second summer crop is grown later. One hazard, particularly in the spring when there is a scarcity of other greenery, is that cottontail rabbits, desert jackrabbits, and cattle prefer tepary vines over other crop plants. Insect pests such as the red spider mite, white flies, grasshoppers, and black cutworms can be a problem, and no pesticides have been approved by the EPA for protection of tepary beans. They do not appear to be harmed by cotton-feeding pests such as the stinkbug or leafhopper. Weeds are kept under control by hoeing or cultivating under dry-farming conditions. The ratio of beans to foliage is higher on a big watered tepary plant on good soil than one grown under marginal conditions. Teparies are not well adapted to cool and humid climates. For example, in field trials in Minnesota the yield was 1,108 pounds per acre when irrigated and 1,990 pounds per acre unirrigated.

Nutritionally the tepary bean compares favorably with other more common grain legumes. The crude protein content of both white and brown domesticated teparies ranges from about 23 to 25 percent, and the value is similar for the wild variety. The protein quality in terms of digestibility and amino acid composition resembles that of common beans. As is true of legumes in general, tepary beans are deficient in tryptophan and the sulfur-containing amino acids, methionine and cysteine, but are a good source of lysine. Since cereal proteins are usually deficient in lysine, but contain adequate levels of cysteine and methionine, the southwestern Indian populations circumvented a potentially serious nutritional problem by the addition of corn, squash, and other vegetables to their bean diet when meat protein was difficult to obtain.

With regard to carbohydrate content, tepary beans contain about 41 percent starch with an amylose/amylopectin ratio of 30/70. *In vivo* digestibility studies have not been reported, but it is assumed that tepary starch is readily digestible when properly cooked. Legume seeds typically contain indigestible sugars such as raffinose, stachyose, and verbascose which cause flatulence. Teparies are no exception, although wild teparies were reported to be very low (1.15 percent) in flatulent sugar levels. The domesticated teparies analyzed were found to contain higher levels (5–8 percent), mainly stachyose, typical of most *Phaseolus* species. Fiber content of teparies is comparatively high, around 6.5 percent. The crude fat content is reported to vary from 0.4 to 2.4 percent with an average of 70 percent being unsaturated. Tepary beans are a good source of minerals, with high values of potassium, calcium, and magnesium. Data on vitamin content are incomplete, but teparies appear to be a good source of niacin, a fair source of riboflavin, and a poor source of thiamine. Three of the bean species native to the American southwest are shown in figure 4.10.

The value of the tepary bean is not that it represents a food with

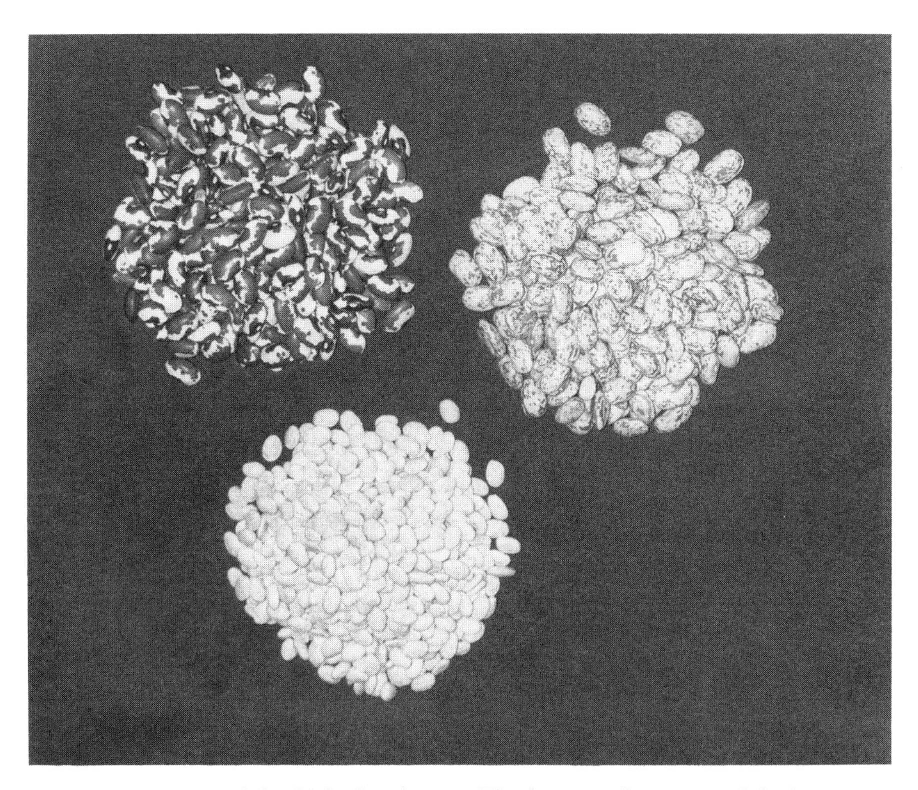

Fig. 4.10. Beans of the U.S. Southwest. The bean at the upper right is the tepary, favored by the Tohono O'odham (Papago) Indians of southern Arizona. At the lower right are Anasazi (a-na-saw-zee) meaning "ancient one" in Navajo, the Indians who still grow them extensively on the high desert of northern Arizona. The mottled at lower left are pinto beans, the most widely eaten bean of the American Southwest.

far superior nutritional qualities, but rather that it is able to produce a crop under conditions which cause more familiar legume seeds to fail. The main advantage of the tepary bean is that of drought avoidance by early maturation of a seed crop (60–70 days). This characteristic, along with the traditional floodwater farming technique developed by the southwestern Indians, permits the production of a crop in most seasons without the aid of conventional irrigation or groundwater use. The National Academy of Sciences has recommended teparies for introduction in the arid and semiarid areas of Africa, South America, Asia, the Middle East, and arid islands in the Pacific Ocean and the Caribbean Sea. However, one important limitation must be considered: teparies must be thoroughly cooked for good digestibility and they require more cooking time than other beans. Therefore, this property limits their usefulness in fuel-poor areas. Looking to the future, the prospects for transferring the beneficial characteristics of teparies to common beans by gene transfer appear favorable.

The *narrowleaf lupin* (*Lupinus angustifolius*) is another interesting legume which only thirty years ago was a virtually worthless wild plant. It may well become the first major field crop domesticated for human food in modern times. Lupin seeds look like white peas and have been used to feed livestock, but they had not been satisfactory for human use because they contained a bitter-tasting alkaloid. This situation changed largely through the remarkable efforts of a western Australian scientist named John S. Gladstones. He sorted through millions of lupin plants in search of low-alkaloid varieties with "sweet" seeds. After a twenty-year search, he was successful not only in finding good-tasting varieties, but types which mature early and have nonshattering seed heads. This provided the first narrowleaf lupin suitable for large-scale commercial production. Western Australian farmers currently grow narrowleaf lupin in many hundreds of thousands of acres and in 1984 harvested some 500,000 tons of the white lupin seeds. Health food stores in the U.S. now offer lupini pasta, a high-protein, fiber-rich pasta made from sweet lupin and triticale flours.

Buffalo Gourd

The *buffalo gourd* (*Cucurbita foetidissima*) (figure 4.11) is a member of the squash family and is indigenous to the arid and semiarid regions of western North America. For thousands of years the American Indians utilized various parts of the buffalo gourd for food, laundry soap, shampoo, stain remover and certain medicinal purposes. While the Indians never cultivated the buffalo gourd, the plant may have developed a dependency upon humans because it is found mainly on disturbed soil such as trash heaps, abandoned cultivated fields, pastures, fence rows, and road beds. In 1946 L. C. Curtis reported on the potential of this species based on observations of native colonies: a perennial plant which would grow on wastelands with little rainfall to produce an abundant

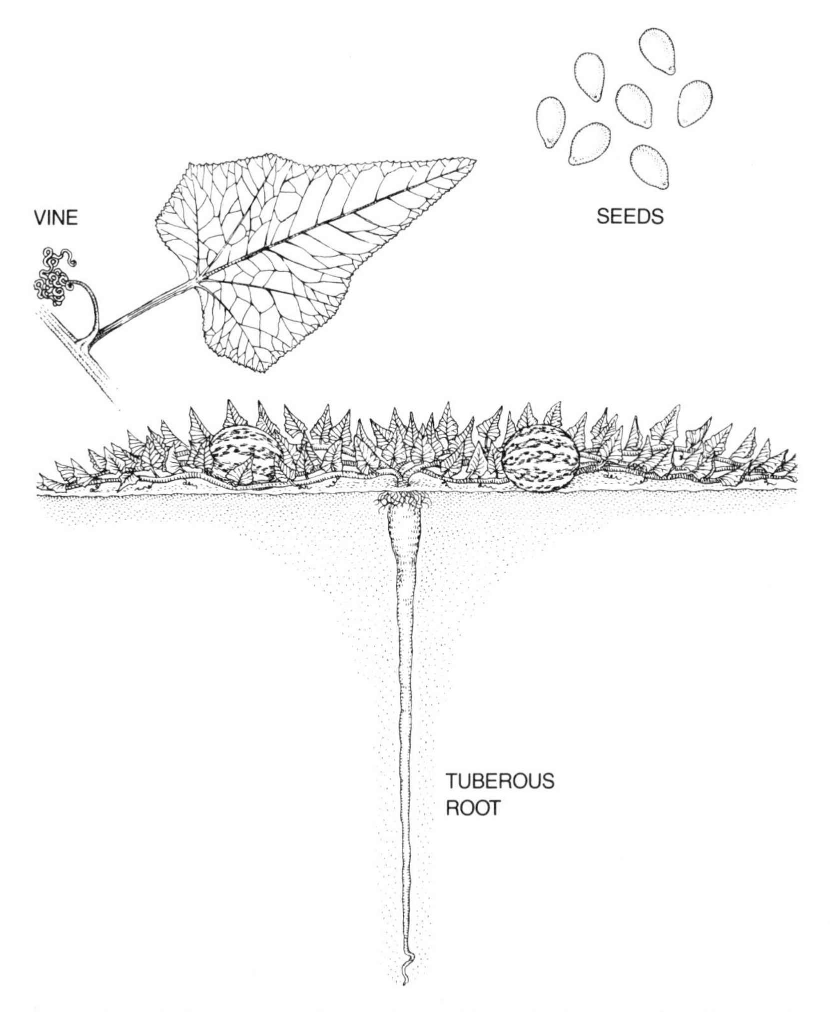

Fig. 4.11. Buffalo gourd (*Cucurbita foetidissima*). Reprinted with permission of *Scientific American.*

crop of fruit containing seeds rich in oil and protein, large roots rich in edible starch, and fruit that could be harvested mechanically. Efforts to domesticate the buffalo gourd were initiated more than twenty years ago. In 1974 Curtis and his co-workers reported the results of a six-year domestication effort carried out in Tel Amara, Lebanon under the sponsorship of the Ford Foundation, using germ plasm from a single collection site in Texas. The project was cut short, however, by the start of the Lebanese civil war. The most significant research on the domestication and utilization of buffalo gourd has been conducted at the University of Arizona for over twenty years by W. P. Bemis and his colleagues, in a multidisciplinary project which included research on agronomy, biochemistry, genetics, and nutrition.

One objective was to develop a homogeneous seed source. Seeds for germ plasm nurseries were collected from a wide geographic area to insure a diverse genetic pool. By 1979 three open-pollinated selections and seven hybrids were tested in a replicated trial to evaluate fruit, seed, and oil yields. Two cultural systems for buffalo gourd production were developed: a perennial mode to allow for optimum seed yield and an annual mode to optimize root yield. Continuing research at the University of Arizona includes hybridization studies, disease reactions, salt tolerance, and nutritional and biochemical studies for food and feed applications. A photo of a domestic field of buffalo gourd is shown in figure 4.12. Photos of the buffalo gourd flower, the mature plant with fruit, and the fruit at various stages of development are given in figures 4.13, 4.14, and 4.15, respectively.

In the wild, buffalo gourd is a perennial because the large, fleshy storage roots survive while the prolific vine growth is killed by sub-freezing temperatures. The primary mode of reproduction is asexual. Large colonies of plants are produced through the development of adventitious roots at the nodes of the vines. Large unisexual flowers occur at most nodes with the predominant sex expression being monoecious, i.e., male and female flowers on the same plant. However, gynoecious, or all female, plants occur widely in native populations as the result of a dominant mutant gene causing abortion of male flower buds. The gourds are usually round, two to three inches in diameter with as many as 200 on a single plant. Each fruit (gourd) contains from 200 to 300 seeds. The seeds weigh about 40 mg. each and contain 30–40 percent edible

FACING PAGE:
Top: Fig. 4.12. Buffalo gourd (*Cucurbita foetidissima*). Germ Plasm Nursery, Tucson, Arizona. Picture courtesy of Office of Arid Lands Studies, University of Arizona.
Bottom: Fig. 4.13. Buffalo gourd (*Cucurbita foetidissima*) pistillate or gynoecious flower. Photo courtesy of Office of Arid Lands Studies, University of Arizona.

Top: Fig. 4.14. Wild buffalo gourd (*Cucurbita foetidissima*) bearing mature fruit. Picture courtesy of Dr. Jess Martineau.
Bottom: Fig. 4.15. Buffalo gourd (*Cucurbita foetidissima*) fruit in various stages of development. Gourd at far right is fully ripe. The one next to it is immature and the two at far left are intermediate to mature, but not ripe.

oil and 30–35 percent protein. The roots, which can attain a weight of up to 80 pounds in three or four growing seasons, contain edible starch similar to that obtained from corn. The extremely prolific vine growth is a ground cover rather than a climber with harsh, sandpaper-like leaves about 8 inches long and with tremendous photosynthetic capability. With the exception of the seeds, all parts of the plant are extremely bitter because of the presence of water-soluble glycosides called cucurbitacins. Studies on water requirements are still in progress, but a total of 17–18 inches during the growing season produced good fruit and root yields.

In the Arizona studies, seed yields using the perennial mode of production have been in the 2,000–3,000 pounds per acre range. The oil content of the seeds from seven hybrids was 38 to 41 weight percent with a mean value of 39 weight percent. The mean fatty acid composition of the triglycerides from the seven hybrids was as follows: linoleic, 61.5 percent; oleic, 27.1 percent; stearic, 3.6 percent, and palmitic, 7.8 percent. This composition is similar to that of the best high-grade edible vegetable oils such as sunflower, corn, peanut, and safflower oils. Linoleic acid is a dienoic (two double bonds) unsaturated fatty acid classified as an essential nutrient for human beings. Oleic acid is the monounsaturated fatty acid so prominent in olive oil and highly regarded by nutritionists. Buffalo gourd oil is similar to soybean oil except that the latter contains a significant amount (about 11 percent) of linolenic acid, a conjugated trienoic acid which is prone to oxidation resulting in objectionable taste and smell. All the findings indicate that buffalo gourd oil would provide a valuable new polyunsaturated food oil. Yields of this oil from buffalo gourd grown in perennial production are better than 90 gallons per acre, compared to about 47 gallons per acre for both sunflower and soybean oils grown on the high plains of Texas. On the basis of oil production alone, buffalo gourd would appear competitive with these two established crops of the high plains.

In addition to the oil, some 30 percent of the weight of buffalo gourd seed is protein with an amino acid composition similar to that of many oilseeds. It is somewhat deficient in threonine, methionine, and lysine (55 percent, 57 percent, and 70 percent of requirement, respectively). Soybean protein is the exception in that it contains significantly more lysine. As would be expected, buffalo gourd seed protein requires amino acid supplementation or blending with complementary protein for complete nutrition in monogastric animals. Current information indicates the seed protein yield to be about 530 pounds per acre. This is more than the yield of seed protein from sunflowers, but less than that from soybeans.

Buffalo gourd root starch is potentially of commercial quality and quantity, offering possible food, fuel alcohol, and other uses (consideration of the fuel alcohol and industrial uses is discussed in chapter 5). The starch content of the root varies seasonally, with a high of 50 percent or better in mid-August. When the vines senesce in the fall the starch percentage declines, but because a major portion of root growth occurs in late summer, a late harvest provides the maximum yield.

Buffalo gourd root is extremely bitter due to its cucurbitacin content. Since these saponins are water-soluble and starch is not, dispersal in water followed by filtration or centrifugation renders the starch essentially free of cucurbitacins. Except that the granules of buffalo gourd starch are much smaller, it is very similar to corn starch. Initial pasting temperature and gelatinization temperature are similar to those of corn starch of the same concentration. Puddings made with buffalo gourd root starch are indistinguishable from those made of corn starch. However, both are considered inferior to puddings made with commercial modified tapioca starch. This suggests that chemical modification to produce a smoother consistency might make buffalo gourd starch more competitive. The protein, fat, fiber, and mineral content of buffalo gourd starch is typical of starches in general and compares favorably as a food source with other common starches.

Halophytes

The planet earth is a very watery place; water covers 71 percent of the earth's surface and the oceans contain some 330 million cubic miles of water, not counting what is locked in the ice caps. Unfortunately, over 99.5 percent of this water is too salty to use in conventional agriculture or is unavailable in ice caps. Most of this salt water is in the oceans, but vast aquifers of saline groundwater exist in many areas. Only one half of one per cent of all the water in the world is available as liquid fresh water. Less than one-third of the earth's surface is land and only about 10 percent of the landmass is suitable for raising crops to feed the world. As pointed out earlier, this valuable cropland resource is shrinking annually for a variety of reasons including salt accumulation wherever irrigation is used to enhance crop production. It is clear, therefore, that the potential for feeding the world's expanding population is shrinking. And the ominous threat of changes in weather and rainfall patterns as a result of the greenhouse effect and depletion of the ozone layer is anything but comforting.

If we had food-producing plants which would thrive on salt water, many millions of acres of land now barren could be made productive. A small group of dedicated scientists has been investigating this possibility and their findings offer great promise for arid-land agriculture. In chapter 3 we discussed efforts made, with some success, to develop salt-resistant varieties of conventional plants by selective breeding. Most conventional crops have dramatically reduced productivity even if they have been bred to tolerate high salinities. Scientists at the University of Arizona, the University of Delaware, and Ben-Gurion University have taken a different approach. Rather than try to alter conventional crops to tolerate high salinity, these researchers have searched worldwide for true halophytes with food and forage potential. They have found a number of plants which not only thrive in salty water, but actually require high salinity.

Pickle Weed

The Environmental Research Laboratory of the University of Arizona has what is believed to be one of the world's largest collection of halophyte germ plasm with more than a thousand acquisitions. They are engaged in the accelerated domestication of two of the more promising halophytes, one of which is a *Salicornia* spp. called pickle weed (figure 4.16), and code-named SOS-7, which produces high-grade fodder and an edible oil seed. Carl N. Hodges and his co-workers have been growing this selection for several years on demonstration farms in Mexico and in the United Arab Emirates. It is being commercialized by a new company named Halophyte Enterprises.

SOS-7 can be cultivated using conventional farming practices and machinery to yield in a seven-month growing season about 8 metric tons per acre of whole dried plant material. This yield is obtained by irrigating with nothing but sea water with salinity levels of 36,000 to 42,000 ppm. This halophyte's minimum salt requirement is believed to be about 5,000 ppm and fresh water will

Fig. 4.16. Fields of *Salicornia* spp. (pickle weed) near Kino Bay, Sonora, Mexico. Pictured in field, Dr. Jess Martineau, agronomist/plant breeder, Native Plants, Inc., Salt Lake City, Utah.

cause it to wilt. The yield compares favorably with that provided by alfalfa, an established forage crop which yields 2 to 8 metric tons per acre, but requires large quantities of fresh water.

About 10 percent by weight of the SOS-7 crop is oilseed, which contains no salt. Some 30 percent of the seed is a high-grade, polyunsaturated vegetable oil very similar to safflower oil. The meal left after removal of the oil is 43 percent protein suitable as feed for livestock and poultry. The straw remaining after removal of the seed contains almost 40 percent salt. Leaching with sea water can remove 65 percent of this salt and a brief rinsing with water of low salt content provides roughage suitable for livestock.

The oil is extracted from the seed by processes similar to those used for soybean oil extraction, but there are unsolved problems associated with it. However, in desert regions where virtually all forage must be imported, the whole plant can be baled and fed to livestock. In such regions, SOS-7 is more valuable as an animal feed than as a source of edible vegetable oil.

The specific requirements to grow SOS-7 make it ideal for most arid regions: air temperature must be constantly above 70 degrees F. during the last 120 days of its growing season and the salt water used for irrigation must be above 65 degrees F. Although sea water or saline aquifer water and soil provide most of the nutrients required, some additional fertilizer will increase the yield. According to Hodges, the design, construction, operation, yield, and revenue of a halophyte farm are comparable to those of a well-managed conventional farm in the same region. The big advantage is that halophytes require little or no fresh water and there are hundreds of thousands of square miles of coastal desert land and land over saline aquifers suitable for halophyte farming.

WildWheat Grain

Certain halophytes of the genus *Distichlis* offer even more exciting potential and the story of their discovery and development is fascinating. Actually it was rediscovery, because in 1885 Edward Palmer described a grain-producing salt grass (*Distichlis palmeri*) which the Cocopah Indians gathered over an estimated 40,000 to 50,000 acres near the mouth of the Colorado River. The plants were watered by the river and salty tidal waters, but when upstream dams began tampering with the flow of the Colorado River, the wild salt grass population declined until by the mid-twentieth century the plant was thought to be extinct.

As mentioned earlier, scientists associated with the Environmental Research Laboratory in Tucson, Arizona conducted a worldwide search for particularly useful halophytes. Some of the most promising were found close by on the coast of the Gulf of California. A small wild population of *Distichlis palmeri* was found on the northern reaches of the Gulf of California. When preliminary studies on growing the salt grass at the Environmental Research Laboratory of the University of Arizona test plot yielded only the equivalent of about one pound per acre, the study was abandoned. But Nicholas Yensen, a young ecologist who had been laid off by the laboratory, and his wife Susana, refused to give up. They formed a small organization (which eventually grew into NyPa, Inc.) and continued the study in their own backyard. With

Fig. 4.17. Distichlis palmeri (Nypa WildWheat Grain) freshly cut seed heads. Photo courtesy of NPY-Nypa, Inc.

assistance from the Tinker Foundation and the Charles A. Lindbergh Fund they were able to continue to explore for additional germ plasm. Their persistence and hard work paid off. After some 16 years of studying the plant and selecting and hybridizing strains, determining the optimum salt levels, and experimenting with methods of growing salt grass, they developed several halophytes with outstanding characteristics.

NyPa (pronounced nee-pah) is the Cocopah Indian name for a rare *Distichlis* grain. To launch the commercialization of the *Distichlis* spp., a corporation named NyPa was formed. In addition to the cereal grain, this organization holds patents on three other unique halophytes which they plan to market worldwide.

The grain variety of *Distichlis palmeri* (see figures 4.17 and 4.18) resembles wheat in much the same way that wild rice resembles rice. So the term *WildWheat* grain was coined and trademarked.

Fig. 4.18. Nypa WildWheat Grain (*Distichlis palmeri*). From this picture one can gain a perspective of the size of the plant and the grain it produces. Photo courtesy of NPY-Nypa, Inc.

This grain possesses a number of remarkable qualities. Even when grown in full-strength sea water, the grain is not salty and has a total ash content lower than that of wheat or barley. This is because the plant actually excretes salts via salt glands, so even the foliage can be grazed by cattle with no ill effects. Ordinary wheat and other cereals contain phytates which can bind essential minerals, causing them to be excreted rather than absorbed. WildWheat grain contains very little of this anti-nutritional factor, but is unusually rich in bran and fiber. Forty per cent by weight is bran and the fiber content is 8 percent compared with 3 percent for regular wheat. With regard to protein content, WildWheat has a lower content than ordinary wheat, 8.7 percent vs. 8–16 percent, but its protein contains generous amounts of all the essential amino acids with the exception of the sulfur-containing amino acids, methionine and cysteine. The starch content, 79.5 percent, is about the same as regular wheat. Two qualities, markedly different from regular wheat, are that *Distichlis palmeri* grain contains no gluten, the principal cause of wheat allergies, and this remarkable grain is a perennial. This means that it does not require yearly planting, a significant economic advantage and important with regard to erosion control. Straw from wheat and other common grains provides little or no nutrition and is useful mainly as bedding or mulch, but the hay from WildWheat grain may have sufficient nutritional value to serve as a forage as well. *Distichlis* spp. are incredibly adaptable. They are perennial salt grasses occurring in the wild in intertidal zones from Canada to Argentina, and so versatile that they can grow in fresh water as well as saline water up to 40,000 ppm. Some types grow well at temperatures ranging from below freezing to 122 degrees F. In their studies, the Yensens found that virtually all of the potential crop species grew better at salt levels commonly found in salinized agricultural areas than those found along desert coasts. This is a decided advantage because inland there are some 6 million square miles of salty land, whereas there are only about 20,000 miles of desert coast land. With so much salt-ruined land to work with, the Yensens encountered no difficulty in enlisting the help of local farmers to provide test plots, and by 1986 the fields in northwest Mexico produced a ton of grain. The yields in the larger test plots were 20 to 30 bushels per acre, which is about the same as for dry-land wheat production. However, yields in some of the small-scale experiments indicate that some strains offer the potential for yields of at least 80 to 100 bushels per acre.

Domestication and development of a new crop is a difficult and challenging task. With domestication well along and with supplies of grain mounting, it was time to pursue the development and commercialization of WildWheat grain. Again, using the wild rice example as a model, it was recognized that introduction of a new food has the best chance of acceptance if it is presented as a gourmet item, so the Yensens initiated marketing through the Neiman-Marcus store in Dallas and in selected restaurants in Tucson.

NyPa Forage

The second *Distichlis* selection for development is a forage variety which will produce as well as alfalfa in soils with twice the

Fig. 4.19. Nypa Forage (*Distichlis palmeri*). A field being cut for hay in Mexico. Photo courtesy of NYP-Nypa, Inc.

maximum salinity that would kill alfalfa. The hay of NyPa forage (figure 4.19) is as nutritious for livestock as alfalfa. It can produce 3–4 tons dry weight per acre per cutting with four or more cuttings per year. In addition, this selection requires less volume of water than alfalfa. In contrast to some other salt-tolerant forages, there is no accumulation of salt in the tissues of this *Distichlis* spp. Grazing or frequent mowing is possible. Field studies with cattle indicate that heavy grazing encourages high protein levels of 12–25 percent in new leaves. New fields are established using tillers (rhizomes). This forage species will grow in virtually all soil types, but does best in heavy soils utilizing water with 2,000 to 15,000 ppm salt. Sea water can be used, but better performance is obtained with lower salinities. Depending on the initial planting density, 3–6 months are required for total coverage of new growth.

Since this species, like WildWheat grain, is a robust perennial, replanting would not be required for at least five years.

This work by the Yensens and their co-workers inspires two profound çonsiderations: 1) the enormous potential for these plants that thrive on salty lands will be of monumental importance to the world, and 2) it emphasizes the urgent need to protect virgin saline habitats to prevent the destruction of coastal salt-tolerant plant species before they are evaluated scientifically.

One *Distichlis* variety is being grown in tissue culture in the Department of Biochemistry at the University of Arizona. The objective of this biochemical and molecular biology research is to understand the mechanism of salt tolerance, which enables this plant to thrive under conditions that are deadly to conventional crops. They plan to study gene transfer and expression into species that are not salt-tolerant.

Salt Bushes

The genus *Atriplex* consists of about 200 species of halophytes, many of which offer potential as forage crops. Recent studies indicate that several of these salt bushes may adapt readily to routine agronomic production. These species offer arid-land plants which can be irrigated with saline or brackish water to produce plants with high biomass, high protein, and mineral levels adequate for animal nutrition. They represent additional candidates for expansion of agricultural production into areas unsuited for traditional agriculture. Most of these perennial salt bushes possess the C4 metabolism which provides the outstanding photosynthetic efficiency expected of plants which thrive under conditions of water stress, high light intensity, and high temperatures.

During the early part of the twentieth century, at least five *Atriplex* species were recognized as useful forage crops for regions subject to summer drought. These plants are capable of providing forage when other summer feed is scarce. They have root systems which penetrate deeply to reach subsoil moisture that accumulates during the winter. They provide high contents of digestible protein, calcium, and phosphorus, and several species produce considerably higher yields than alfalfa. The perennial evergreen species of *Atriplex* can serve as a source of forage during the winter months in cold climates. Two species of *Atriplex* are pictured in figures 4.20 and 4.21.

Recent studies indicate that these salt bushes manage to cope

Fig. 4.20. A plantaion of *Atriplex* spp. established near Beersheva, Israel, to provide forage for livestock. Rainfall in this location is about 8 inches annually. At the left is *Atriplex nummularia*; in the center (heavy with fruit) is *Atriplex canescens*. Atriplexes can make areas productive that are otherwise useless and salt-devastated. Note the barrenness of the surrounding landscape (Photo by M. Forti).

with high levels of salinity by adjusting osmotically and by gradually accumulating salts in giant vesiculated trichomes on the surface of the leaves. When excess sodium ion is accumulated, ionic balance is achieved by synthesis of organic acids, such as oxalic acid, in the bladder cells. This allows the solute potentials in the remainder of the plant to remain at moderate levels, permitting normal metabolism to proceed. Eventually the vesiculated hairs (trichomes) which collect the salts burst and leave a litter of salt crystals and cell wall debris on the surface of the leaf. *Atriplex* species also manage to cope with specific ions which are often toxic to other plants. For example, *A. polycarpa* has been shown to tolerate 39,000 ppm of sodium chloride and 80 ppm of boron. These crops might be grown and harvested simply to remove excess salts and toxic elements to reclaim land for conventional agricultural production.

With regard to productivity, three *Atriplex* species indigenous to

Fig. 4.21. Awassi ram grazing on *Atriplex canescens* during the height of summer. This species has proven to be one of the most palatable in trials in Israel (Photo by M. Forti).

the western United States have been grown under routine agronomic conditions with limited irrigation to produce 8,000–9,000 lbs. per acre per year of biomass on a sustained basis. Even though *Atriplex* species grow well in desert soils deficient of nutrients, all species tested responded favorably to fertilization. It has been observed in these studies that individual plants achieved exceptional biomass productivity. This suggests that selection for high-yielding genotypes is possible. Most of the perennial forage species of *Atriplex* can be propagated by direct seeding, vegetatively by the use of cuttings, or by layering. Even though the mature plants have great salinity tolerance, seedlings have difficulty coping with salinity, so at the time of germination fresh water (usually rainfall) is important to dilute the salts during the early establishment phases. Weed control can be a serious problem in *Atriplex* establishment, but cultivation is the most effective method. *Kochia* and *Amaranthus* species are important weeds for competition with *Atriplex* in the southwest U.S. However, recent studies suggest that

Fig. 4.22. Scientists in western Egypt examining *Atriplex nummularia,* a forage crop more appropriate than grass or alfalfa for arid regions. Photo by David L. Moore.

these species may also be worthwhile forage crops and the possibility of growing two or more of these species together should be investigated. There has been some concern that oxalic acid accumulation in certain halophytes could be toxic to livestock, but recent studies with *Kochia* and various species of *Atriplex* indicate that oxalate levels are not high enough to cause toxicity problems in cattle and sheep.

Recent investigations in Egypt and Israel have shown *Atriplex nummularia* (figure 4.22) to be superior to any other of the species studied in terms of biomass, growth rate, resistance to overgrazing, etc. Dr. H. N. Le Houérou has spent twenty-five years studying sheep nutrition in northern Africa and concludes that sheep can be raised very successfully on a diet consisting entirely of cactus and salt bush. He studied ten shrubs and some thirty mixtures and found that the best performance was achieved with a combination of *A. nummularia* and *Opuntia ficus-indica* in equal amounts. He reported that sheep fed entirely on *Atriplex* consumed up to 4.4 pounds dry weight daily, but drank up to 1.85 gallons of water. The same sheep on the combination diet consumed only 0.66 gallons per day, the same as sheep on a non-halophytic diet. These findings indicate the potential benefits of raising *Opuntia ficus-indica* and *Atriplex nummularia* in arid regions to provide a forage crop more acceptable than grass or alfalfa for these regions.

Sordan grass is another interesting halophyte which is the result of crossing sorghum with Sudan grass. This plant is capable of producing good forage on marginal soils and at the same time reclaim some damaged soils so they can again be used to produce plants for human food. Sordan grass repairs sodic soils by releasing acids which dissolve calcium carbonate to release calcium ion and carbon dioxide. The calcium displaces the sodium in the sodic clay soil and it combines with the carbon dioxide to form sodium bicarbonate, which is water-soluble and washes away. *Tecticornia* is a salt-tolerant plant wich grows in the coastal mud flats of Australia. The Australian aborigines grind the seeds of this halophyte into a nutritious flour to be used in making bread. The National Research Council recently published a report listing hundreds of salt-tolerant plants, including grasses, shrubs and trees.

Spirulina

When the Spanish conquistadores arrived in Tenochtitlan, the present site of Mexico City, they found the Aztecs harvesting *Spirulina geitleri* which grew wild in Lake Texcoco. *Spirulina* are algae which grow naturally in highly saline and alkaline lakes in hot climates. Depending on the conditions of growth, *Spirulina* can contain 60 percent or more of high-quality protein with all of the essential amino acids required in human nutrition. The algae also provide important unsaturated fatty acids. This was beneficial to the Indians of Tenochtitlan because their other major source of protein was corn, which is deficient in several essential amino acids. Today in Africa, some of the nomadic people of Chad skim an indigenous *Spirulina* alga off Lake Chad, whereupon it is dried and eaten by the local population.

Algae cultivation offers several potential advantages in addition to providing high-quality protein. In arid climates it can be grown in ponds or tanks of brackish water on a continuous year-round basis utilizing the abundant solar radiation available in these regions. While its use as a human health food probably represents its highest economic value, it offers a very nutritious feed ingredient for livestock, fish, and other aquatic organisms. It could serve as a source of beta-carotene, carrageenin, tocopherols, and could be used as a feedstock for ethanol fermentation.

Currently, at the New Mexico Solar Energy Institute of New Mexico State University, Barry Goldstein and his co-workers are conducting a project to determine the feasibility of growing *Spirulina* on a commercial basis using saline groundwater. New Mexico has some 15 billion acre-feet of saline groundwater available with salinities too high for growing traditional crops. These researchers consider that the feasibility of cultivating *Spirulina* in New Mexico has been demonstrated, and now they are turning their attention to market research. *Spirulina,* mainly in the form of compressed tablets or in capsules, is available currently as a food supplement in health food stores.

Quality-Protein Maize

Several hundred million people in Latin America, Africa, and Asia depend on maize (corn) for their daily food, and for many it is their main source of dietary protein. Traditional maize varieties are poor in nutritional quality because they are deficient in the essential amino acids lysine and tryptophan and the vitamin niacin. As a result, traditional maize, without supplementation, cannot sustain acceptable growth and satisfactory health, especially in children, pregnant and lactating women, and the sick. For many years researchers attempted to correct this deficiency by breeding nutritionally improved varieties of maize. Finally, in 1963, scientists at Purdue University found a mutant maize that had about twice the usual levels of lysine and tryptophan and a nutritive value about 90 percent of that of skim milk protein, the standard against which cereal protein is measured. This discovery was greeted enthusiastically. It was estimated that this "new" maize would add ten million tons of quality protein to the world food supply and alleviate widespread malnutrition.

The new maize was called "high-lysine" maize by some, but this was a partial misnomer, because it is also high in tryptophan and its nutritional benefits exceed those provided by lysine alone. Opaque-2 was a more appropriate name because the mutant maize produced soft, opaque kernels instead of the hard, transparent kernels typical of most varieties. It was designated "2" because it was the second mutant the researchers discovered in the opaque group. Maize breeders around the world began transferring opaque-2 genes into local maize varieties and the new crop was rushed into production. In a short time opaque-2 varieties were being marketed in South America and research programs were underway in Eastern Europe and the Soviet Union. In the U.S. production rose from practically nothing in 1970 to an estimated 240,000 tons in 1975.

Results of nutritional studies with opaque-2 maize have been encouraging; significant improvement in protein deficiency malnutrition and prevention of pellegra have been reported. Unfortunately, the undesirable characteristics of opaque-2 caused it to be discredited. The grain was chalky, not shiny as preferred in most regions. The ears were small, yields were 8–15 percent lower than those of traditional varieties, it dried slowly, and was more susceptible to damage by fungi and insects. Farmers refused to grow it because of its poor field performance and millers were reluctant to handle it because of its poor storage characteristics. Consumers didn't like it because of its soft, floury texture and unconventional appearance.

By the late 1970s most research organizations had abandoned opaque-2, but breeders and chemists at the Centro Internacional de Mejoramiento de Maiz y Trigo (CIMMYT) near Mexico City tried one approach after another to improve opaque-2. They concentrated their efforts on genetic modifications to harden kernels, to raise yields, to make the endosperm transparent, and to make the grain dry faster. After ten years, persistent effort brought about continuous improvement in qualities to raise farmer and consumer acceptance without losing the desirable nutritional qualities. This was made possible to a large extent by sensitive analytical methods developed by Evangilina Villegas, enabling them to sample a single kernel without damaging its ability to germinate. This research produced a new class of maize that combined the nutritional advantages of opaque-2 with the kernel structure of conventional maize varieties. CIMMYT named their improved opaque-2 "quality-protein maize" (QPM).

Low grain yield was probably the most damaging characteristic

Fig. 4.23. Ears of QPM (right, NUTRICTA) are indistinguishable now from those of normal maize (left, ICTA). Photo kindly provided by National Research Council.

of opaque-2. Today, QPM genotypes can equal the yields of conventional maize varieties under cultivation in developing countries. In fact, several experimental QPM varieties outperformed the normal maize in several regions of the world. QPM kernels are shiny, transparent, and just as hard as those of traditional maize, so kernel density and texture are no longer factors restricting acceptance by farmers, millers, and consumers. Today's QPM dries at a rate comparable to that of normal maize, so drying time has ceased to be a barrier. Figure 4.23 shows how ears of QPM are indistinguishable from those of conventional maize. QPM varieties are still more susceptible to ear rot than normal maize, but this susceptibility is decreasing generation by generation, and other diseases seem to attack QPM and normal maize with comparable severity. With regard to nutritional qualities, the QPM of today retains all of the advantages of the opaque-2 of the 1970s.

QPM clearly holds out outstanding promise, especially for countries with high instance of malnutrition. Any lingering doubts about its potential need to be quickly resolved by an organized, international cooperative research effort in order to provide the steady and orderly development of what appears to be an important new food crop. In some parts of the world QPM could be used to advantage instead of cassava, plantain, sorghum, or other inferior-protein staples.

QPM is not yet ready for large-scale use in the U.S. because it is available only in open-pollinated types rather than the hybrids mainly used in North America. Research is needed to tailor QPM to U.S. needs and efforts are underway with promising early results. For arid lands there are drought-resistant strains of maize, such as the Hopi blue corns, which have been grown in the American Southwest for more than a millennium. It is hoped that researchers will be able to introduce QPM's superior nutritional qualities into these ancient desert strains to offer an improved variety for arid lands.

Other Possibilities

The examples of lesser-known plants offering potential for use as food crops in dry climates discussed thus far indicate the wealth of possibilities available from nature's storehouse. But there are many more. While it is beyond the scope of this work to present an exhaustive study of the subject, a few additional plants will be mentioned briefly to illustrate just how bountiful nature really is.

Scientists at the University of Sydney in Australia have been investigating the nutritional value of native food plants of the arid and semiarid regions of Australia. These native food plants, some 100 to 150 known edible species, enabled the desert aboriginal people to survive in one of the harshest environments on earth. These recent studies have revealed that seeds of *Acacia, Eragrostic, Fimbristylis, Panicum,* and *Portulaca* species are almost all twice as rich in protein as the common cereals and often many times higher in fat content. The rootstocks of various *Ipomoea, Vigna,* and *Dioscorea* species offer nutrients similar to those of carrot and potato. The fruits of various *Solanum, Santalum* and *Ficus* species have higher protein, fat, and carbohydrate contents than cultivated fruits, and all of these grow under desert conditions.

In the arid and semiarid regions of Africa there are many native

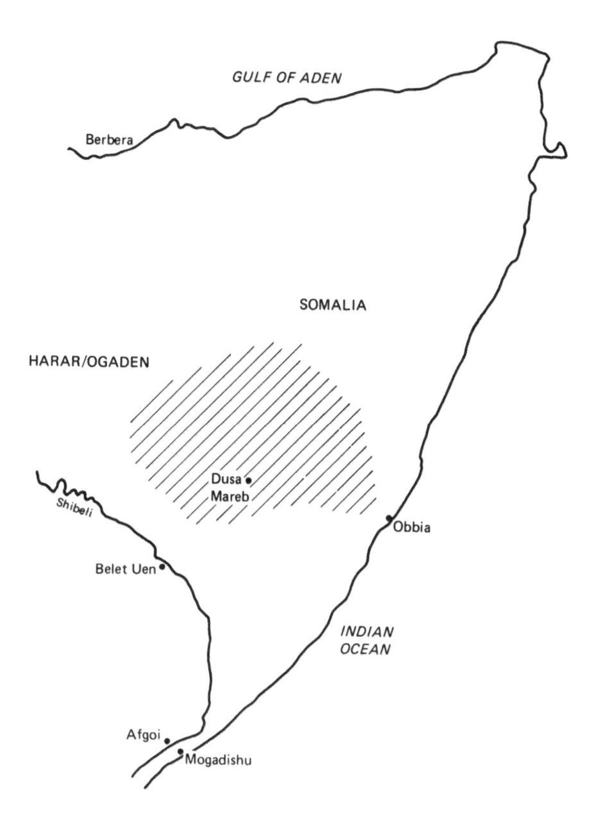

Fig. 4.24. Map showing ye-eb nut native habitat in the horn of Africa. Source: National Research Council.

food plants well-adapted to the desert conditions, but virtually unknown outside the areas where they grow. A recent publication lists 333 species of food plants used by the natives who inhabit the arid and semiarid regions of Namibia, Botswana, and southern Angola. Nutritional data have been determined for 66 of these species and a number are seen as important candidates for domestication. One plant native to the desert regions of Somalia and Ethiopia, the wild legume *ye-eb* (*Cordeauxia edulis*), has been highly recommended for further investigation. Ye-eb is a shrub which produces a chestnut-like fruit similar to macadamia and pistachio nuts. Few other plants can survive in the hot, dry, poor soil of the ye-eb's native habitat. During the recent African droughts, ye-eb was one of the few foods available, so the nomads and their livestock overexploited the native stands, thereby threatening the plant with extinction. Ye-eb deserves protection and cultivation to make sure that it is not lost. It is indigenous in only two very small regions, one of which is indicated by the shaded area on the map in figure 4.24.

Another creature, not a plant, is worthy of mention and recommendation for arid-land agriculture: the one-humped Bedouin camel. This animal, commonly thought of only as a beast of burden, is significantly superior to cattle, sheep, and goats in terms of food production, its influence on the environment, and even in controlling human population growth. The average female camel produces five to ten times more very nutritious milk per lactation period than a cow. One pound per day will provide the entire daily protein requirement for a full-grown man. The milk does not sour even when held at ambient temperature and it contains three times as much vitamin C as cow's milk. The camel thrives on salty, thorny browse and can do well with water available for only one hour every ten days. Furthermore, a lack of water does not affect the volume or quality of milk secreted. Camels eat the leaves of shrubs, trees, and herbs as well as grass when it is available, but they do not overgraze any type of vegetation. They do not trample and compact the soil because they have soft, flat hoofless feet and they are more than four times as efficient as cattle in converting animal forage into human food.

The camel is the only domesticated animal that has demonstrated its ability to influence human population growth. In pastoral societies, if livestock provide the primary source of food, and if they regulate the creation of new families by the availability of the livestock, then in the absence of outside economic inputs, the human population cannot grow faster than the herd. Because of the slow growth of camel herds, societies which depend on them for their livelihood practice various kinds of social controls to regulate marriage and birth. The advantages offered by the camel for the future of pastoral economies in arid and semiarid regions should not be overlooked.

In South America there are more than twenty species of legumes, tubers, grains, and fruit, "the lost crops of the Incas," that should be investigated. In Mexico, just in the Sonoran Desert alone, there are several dozen species including both wild and domesticated food crops, many of which offer agronomic promise. It is abundantly clear that there is no shortage of attractive candidates to "make the desert bloom" without irrigation, but unfortunately there has been a shortage of interest and funding to develop them.

SELECTED INFORMATION SOURCES

Aronson, J. A. 1989. *Haloph: A Data Base of Salt Tolerant Plants of the World.* Tucson: University of Arizona Press.

Bassett, C. A. 1990. "Rebirth for Ancient Seeds." *Arizona Highways* 66(6):36–41.

Bohnert, H. J., and N. Yensen. 1987. "Alternative Crops, Arizona Land and People." *Magazine of the College of Agriculture* 38:10.

Bower, N., K. Hertel, J. Oh, and R. Storey. 1988. "Nutritional Evaluation of Marama Bean (*Tylosema esculentum,* Fabaceae): Analysis of the Seed." *Economic Botany* 42(4):533–540.

Brand, J. C. and V. Cherikoff. 1985. "Australian Aboriginal Bushfoods: Their Nutritive Value." In E. E. Whitehead, C. F. Hutchinson, B. N. Timmermann, and R. G. Varady, eds., *Arid Lands Today and Tomorrow. Proceedings of an International Research and Development Conference, Tucson, Arizona.* Boulder, Colo.: Westview Press.

Cooperative Arid Lands Agriculture Research Program Newsletter. 1988–89.

Crosswhite, F. S., ed. 1983. "A Special Issue on the Tepary Bean." *Desert Plants* 5(1):2–63.

Golstein, B. 1985. "*Spirulina,* A New High-Value, Saline Water Crop for Energy, Food and Feed Production in Arid Lands." In E. E. Whitehead, C. F. Hutchinson, B. N. Timmermann, and R. G. Varady, eds., *Arid Lands Today and Tomorrow. Proceedings of an International Research and Development Conference, Tucson, Arizona.* Boulder, Colo.: Westview Press.

Goodin, J. R. 1979. "*Atriplex* as a Forage Crop for Arid Lands." In G. A. Richie, ed., *New Agricultural Crops, AAAS Selected Symposium,* pp. 89–125. Boulder, Colo.: Westview Press.

Gradus, Y., ed. 1985. *Desert Development—Man and Technology in Sparselands.* Dordrecht: D. Reidel.

Hinman, C. W. 1986. "Potential New Crops." *Scientific American* 255:33–37.

Hodges, C. N., and W. L. Collins. 1984. "Future Food Production: The Potential is Infinite, the Reality May Not Be." *Proceedings of the American Philosophical Society* 128(1):27–30.

Hodges, C. N., W. L. Collins, and J. J. Riley. 1987. "Direct Seawater Irrigation as a Major Food Production Technology for the Middle East."

Presentation to the Center for Strategic and International Studies, Georgetown University, Washington, D.C., June 25.

Hogan, L., and W. P. Bemis. 1983. "Buffalo Gourd and Jojoba: Potential New Crops for Arid Lands." *Advances in Agronomy* 36:317–349.

Le Houérou, H. N. 1986. *Forage and Fuel Production from Salt Affected Wasteland.* Elsevier.

Lochhead, C. 1988. "Tilling a Market for Exotic Products." *Insight* 4(6):46–47.

National Academy of Sciences. 1975. *Underexploited Tropical Plants with Promising Economic Value, BOSTID(JH-217D).* Washington, D.C.: National Research Council.

National Academy of Sciences. 1979. *Tropical Legumes: Resources for the Future, BOSTID (JH-217D).* Washington, D.C.: National Research Council.

National Research Council. 1984. *Amaranth: Modern Prospects for an Ancient Crop.* Washington, D.C.: National Academy Press.

National Research Council. 1988. *Quality-Protein Maize.* Washington, D.C.: National Academy Press.

National Research Council. 1989. *Triticale: A Promising Addition to the World's Cereal Grains.* Washington, D.C.: National Academy Press.

National Research Council. 1990. *Saline Agriculture: Salt Tolerant Crops for Developing Countries.* Washington, D.C.: National Academy Press.

Powell, A. M. 1987. "Marama Bean (*Tylosema esculentum,* Fabaceae), Seed Crop in Texas." *Economic Botany* 41(2):216–220.

Russell, C. E., and P. Felker. 1987. "The Prickly-Pear (*Opuntia* spp., Cactaceae): A Source of Human and Animal Food." *Economic Botany* 41(3):433–445.

Russell, S. A. 1985. "The Invincible Tepary." *Gardens for All News* 8(May):34–36.

Stiles, D. 1987. "Camel vs. Cattle Pastoralism: Stopping Desert Spread." *Desertification Control Bulletin.* United Nations Environment Programme, no. 14, pp. 15–21.

Vietmeyer, N. D. 1986. "Lesser-Known Plants of Potential Use in Agriculture and Forestry." *Science* 232(June 13):1379–1384.

Weber, L. E., C. S. Kauffman, N. N. Bailey, and B. T. Yolak. 1986. *Amaranth Grain Production Guide.* Emmaus, Pa.: Rodale Research Center, Rodale Press.

Wickens, G. E., J. R. Goodin, and D. V. Field, eds. 1985. *Plants for Arid Lands, Proceedings of the Kew International Conference on Economic Plants for Arid Lands, Royal Botanic Gardens, Kew, England, 23-27 July, 1984.* London: George Allen & Unwin.

Williams, A. A. 1987. "Environment and Edible Flora of the Cocopa." *Environment Southwest* (Autumn):22–27.

Wood, R. T. 1985. "Tale of a Food Survivor: Quinoa." *East West Journal* April:63–68.

Yensen, N. P., and S. B. de Yensen. 1987. "Development of a Rare Halophyte Grain: Prospects for Reclamation of Salt-Ruined Lands." *Journal of the Washington Academny of Science* 77(4):209–214.
Yensen, N.P., S. B. de Yensen, and C. W. Weber. 1985. "A Review of *Distichlis* spp. for Production and Nutritional Values." Paper presented at Arid Lands: Today and Tomorrow, Tucson, Arizona.
Yensen, S. B., and C. W. Weber. 1986. "A Research Note. Composition of *Distichlis palmeri* Grain, a Saltgrass." *Journal of Food Science* 51(4):1089–1090.

FIVE

Arid-Land Plants Yielding Industrial Products

Arid and semiarid agricultural areas are generally prone to erosion from wind and heavy rain for they have less biomass cover to protect them. The more denuded the area the more prone to devastation it becomes. Most food crops, aside from fruits and nuts, are annuals and as such leave the land bare for considerable periods of the year, thus compounding the problem. These lands are naturally lacking in soil organic matter and nitrogen—tenfold lower than temperate agricultural soils. Annuals are naturally high-water-use crops and so must be irrigated when grown in arid or semiarid areas to produce good yields. The more irrigation water applied the more likely, or faster, the soil will become salinized. Obviously, growing high-water-use annuals in arid or semiarid areas jeopardizes the soil and is generally self-limiting over the long term.

Although there is a definite need for additional food crops in many arid and semiarid countries some, like areas of the United States, already produce surplus food. In fact, despite the scenes of famine so often projected by the media from Africa, the world at large is presently awash in food, although this condition is likely to change within one to two decades. Politics and lack of appropriate distribution systems are what cause famines today, not the lack of food per se. Often, well-meaning organizations have introduced mesic plant foods and food crops to peoples and places where they cannot be grown successfully, and often at the expense of indigenous food crops. More consideration must be given to developing and promoting arid-land-adapted food crops for localities that still lack food. For nations like the United States that already have surpluses of food, serious consideration should be given to replacing many of the present food and fiber crops grown west of the 100th meridian with plants requiring less water and that produce industrial products. Most of the plants that produce these items of commerce are perennials.

In order to preserve the land and conserve water, more attention must be given to growing perennial, low-water-use crops. Whenever possible, perennial leguminous crops should be grown to add both organic matter and nitrogen to the soil either as a single crop or as an intercropping scheme.

Plants serve as the source of many products that are important to international trade and industry. These products are often produced from perennials and include gums, resins, fibers, rubber, oils, fuels, chemical feedstocks, and medicinal agents. Many of the plants that provide these essential materials thrive in arid and semiarid regions. While botanical sources have been important items of commerce for centuries, the extent of general scientific knowledge and the elucidation of the complex chemical structures of these natural products have increased greatly in the past few decades. Current knowledge is sufficient to guarantee that plants providing industrial products deserve a prominent place in modern agriculture. During the past several decades petroleum has served as a dominant source of many industrial chemicals as well as fuel. Now that this source is dwindling and is an important contributor to the budget deficit in the United States and many other countries, renewable plant sources should gain in importance. In this chapter we shall describe a few plants which offer attractive alternatives.

Gumweeds

The *gumweeds,* which occur naturally in western North America from southern Mexico all the way to Alaska, produce large amounts of aromatic resins that exude to thickly cover the surface of the plant. This sticky exudate gives rise to the common name "gumweed." The hydrophobic resinous coating is believed to reduce water loss and also to repel pests, thus enabling these plants to survive in stressful climates. The gumweeds are members of the Asteraceae family including *Grindelia* and *Chrysothamnus* species. *Grindelia* species produce bicyclic compounds which have chemical and physical properties similar to the tricyclic terpenoids in the wood and gum rosins of industrial importance. *Chrysothamnus* species produce resins, oils, and rubber.

Grindelia

Grindelia resins are clear, non-volatile terpenoid mixtures insoluble in water, but soluble in organic solvents. Current research indicates that these resins may prove to be the ideal substitute for wood and gum rosins. In addition, they may provide completely new resins with components more amenable to chemical modification than those of the wood rosins.

Wood rosin is the hard, brittle resin remaining after oil of terpentine has been distilled from pine stumps. Industry consumes huge quantities of wood and gum rosins to prepare adhesives, paper sizings, synthetic polymers, varnishes, printing inks, soaps, and many other items of industrial importance. Traditionally, wood rosin was obtained from virgin pine stumps, but this source is all but exhausted, while gum rosin is obtained by tapping living trees, an extremely labor-intensive procedure. For these reasons, the U.S. is faced with a shortage of natural high-quality resins, and while the problem may be alleviated somewhat by imports, a renewable domestic source could be of considerable economic importance.

One gumweed, *G. camporum* (figures 5.1–5.4), has been selected for in-depth research and development. This species is an herbaceous resinous perennial native to the Central Valley region of California. It is salt-tolerant and drought-resistant, and has a woody base that produces several erect, open-branched herbaceous stems with yellow flower heads. Some strains are annuals and many have the ability to sprout from the root crown to produce two crops in a single growing season. When harvested before seeds are formed, the plant becomes a perennial, thus permitting harvesting for several years—perhaps up to five—before replanting is necessary. Formulations of *G. camporum* crude resin have

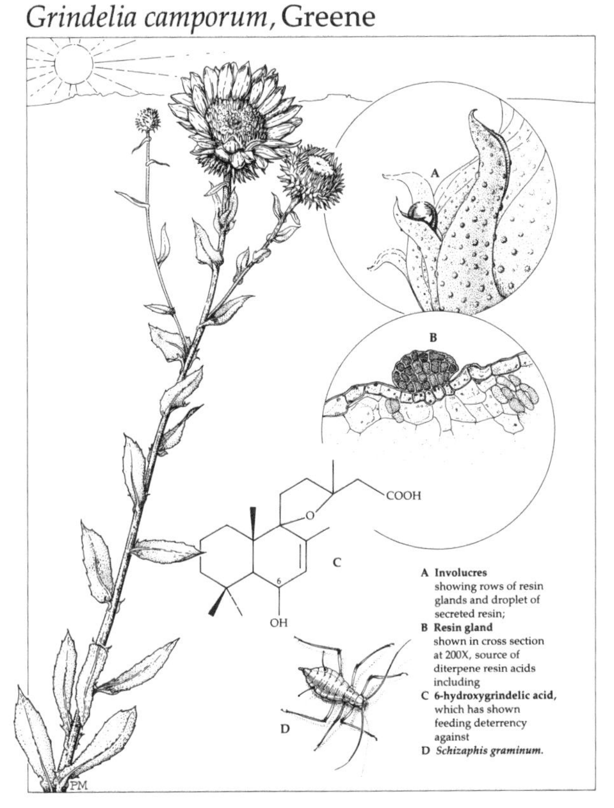

Fig. 5.1. Grindelia camporum. Drawing courtesy of Paul Mirocha of Office of Arid Lands Studies, University of Arizona.

been prepared and tested; the results indicate that the resins could be substituted for wood rosin in many applications and may be superior in some because of greater thermal stability. In addition, *G. camporum* is a potential alternative source of resin acids for use in inks, adhesives, varnishes, printing inks, tackifiers, and other industrial products.

Preliminary economic evaluation suggests a very favorable return on investment by utilizing the resin even as the sole product. The remaining material could be utilized as a fuel, but has greater potential as an animal feed. In addition to the grindelic acid resins, the biomass contains an interesting array of natural products in-

Fig. 5.4. Experimental plot of the gumweed *Grindelia camporum* located at the BioResources Research Facility, Office of Arid Lands Studies, University of Arizona.

FACING PAGE:
Top: Fig. 5.2. The gumweed *Grindelia camporum* just starting to flower. The blossom head and leaves are sticky with exudate. Photo by Dr. Joseph Hoffmann, University of Arizona.
Bottom: Figure 5.3. The gumweed *Grindelia camporum* in full flower. Photo by Dr. Joseph Hoffmann, University of Arizona.

Left: Fig. 5.5. The gumweed rubber rabbitbrush (*Chrysothamnus* spp.) in flower near Ephriam, Utah. Photo by Dr. Cyrus M. McKell.
Right: Fig. 5.6. Flowering rubber rabbitbrush (*Chrysothamnus paniculatis*). Photo by Dr. Joseph Hoffmann, University of Arizona.

cluding polyphenols, alkanes, terpenoids, and steroids. The polyphenols offer possibilities in preparing adhesives and thermoplastics. The alkanes are mostly waxes for which there is a good market. The terpenoids and steroids may be useful in the preparation of pharmaceuticals and agricultural products. Gumweed bagasse contains many potential value-added products and is worth extensive research, both scientifically and commercially.

Rubber Rabbitbrush

Rubber rabbitbrush (*Chrysothamnus* sp.) (figures 5.5–5.7) is a common desert shrub which grows over a wide range of environmental conditions from northern Mexico to southern Canada and from the Great Plains west to the Pacific Ocean. This hardy, wild shrub is of potential value as a source of high-grade natural rubber, as a forage crop for wildlife and livestock, as a landscape plant for arid regions, and as a potential source of hydrocarbons, natural insecticides, and fungicides.

Fig. 5.7. Rubber rabbitbrush (*Chrysothamnus* spp.) growing wild near Ely, Nevada. Photo by Dr. Jess Martineau.

Rabbitbrush is the common name for many *Chrysothamnus* species, five of which are more common than the others: *albicaulis, hololeucus, nauseosus, consimilis,* and *graveolens.* The first three have white or gray stems and mature leaves (young leaves are green) while *C. consimilis* and *C. graveolens* have green stems and leaves. These shrubs ordinarily range in size from 1 to 5 feet in height although 10- to 12-foot species occur. Rabbitbrush thrives under conditions of drought and aridity, nutrient-poor soils, poor soil aeration, winter cold, summer heat, short growing seasons, and high winds, and many species tolerate saline and alkaline conditions. The shrub blossoms in late summer or early fall with spectacular yellow flowers, and is self-fertile with fruits maturing in October and November. The small seeds (some 43,700 per ounce) can germinate almost immediately. Rabbitbrush has a high rate of net photosynthesis for a C3 plant and does not become light-saturated at full sun. The species is widespread and often abundant, but it does not form pure stands except in disturbed areas such as highway shoulders and media.

The earliest indication of the rubber potential for rabbitbrush was in 1878 in St. George, Utah, when some Indians taught several young Mormon boys how to make gum by chewing the inner bark of the shrub. In 1918 and 1919 Hall and Goodspeed published reports on their survey of plants in western North American that offered possibilities as sources of rubber. At the time their study was made, the U.S. was engaged in World War I, and rubber was growing in importance to the mechanized world. Their publications indicated that certain subspecies of *Chrysothamnus nauseosus* could provide natural rubber, and they estimated that over 300 million pounds of rubber was contained in the wild stands of *C. nauseosus* in the western U.S.

After World War I, interest in a domestic source of natural rubber declined until during World War II when the natural rubber from Asia was no longer easily accessible. Once again there was interest in the desert plants containing rubber, and both guayule and rabbitbrush were promoted as possible sources. A government project planted huge acreages of both species, and considerable effort was devoted to selection of high-yielding strains. Unfortunately, as soon as the war ended, the rubber project was terminated after nearly four years of effort. All of the cultivated guayule and rabbitbrush plants were destroyed and all of the high-yield accessions were lost.

Recently, there has been a renewed interest in this area and researchers have begun again to study *C. nauseosus* as a producer of natural rubber and other hydrocarbons. In 1977, Buchanan and his co-workers reported the results of an evaluation of over 100 plants having potential as hydrocarbon and rubber crops. Ultimately they reported data on more than 300 species. They compared guayule and rabbitbrush and concluded that rabbitbrush was better botanically while guayule scored slightly better in hydrocarbon production. In 1980, the National Science Foundation awarded a grant to Native Plants, Inc. of Salt Lake City, Utah to investigate a new source of natural rubber in the U.S. This group, under the direction of W. Kent Ostler, collected data on the rubber content of five subspecies of *C. nauseosus* throughout nine regions of ten western U.S. states. They identified one subspecies, *C. nauseosus viridulus*, with a particularly high rubber content. The average values for rubber in the *viridulus* subspecies were more than double the amounts produced by the next closest subspecies. This suggests that rubber content has strong genetic controls and effort should be directed at selecting high rubber-producing strains.

With regard to environmental stress factors which influence rubber content, they found that rubber concentration was higher in plants growing in saline valley bottoms, on gentle slopes at low elevations, and with water stress. They determined the average growth rate to be 0.75 pound a year, a value in close agreement with that reported by Hall and Goodspeed in 1919. It is interesting that C. M. McKell reported that growth rates of plants that had been clipped or hedged and then allowed to grow for two years yielded average values of over one pound a year. Ostler proposed that in an agricultural setting in which the harvest would take only the above-ground growth and leave the root system intact for re-growth, it would be possible to harvest over 300 pounds of rubber per acre in six years.

The rubber produced by *C. nauseosus* is referred to as Chrysil rubber. Hall evaluated the quality of Chrysil rubber and reported in 1919 that it had mechanical properties similar to those of *Hevea* rubber and that it could be vulcanized. There seemed to be no significant difference between the rubber from guayule, *Hevea,* and rabbitbrush. The recent studies by Ostler and co-workers utilizing the modern techniques of infrared analysis and nuclear magnetic resonance show that Chrysil rubber is essentially a *cis*-1,4-isoprene polymer.

In addition to rubber, *Chrysothamnus* species produce significant amounts of potentially useful resins and oils. Ostler et al. found that environmental factors were not important in influencing the resin content, but, as with rubber production, only the subspecies *viridulus* was significantly better than the other subspecies. This plant is an impressive chemical factory. Preliminary chemical studies indicate that some 500 secondary metabolites such as terpenes, phenols, labdane-type acids, etc. are produced. Some of these compounds are candidates as natural insecticides, fungicides, and insect repellants. The white-stemmed subspecies are attractive as ornamental landscape plants for arid regions.

Rabbitbrush should be carefully researched as a prime candidate as a new commercial crop for semiarid regions. It is widely distributed throughout the western U.S. and grows on marginal land—alkaline and saline soils—which is currently unused or un-

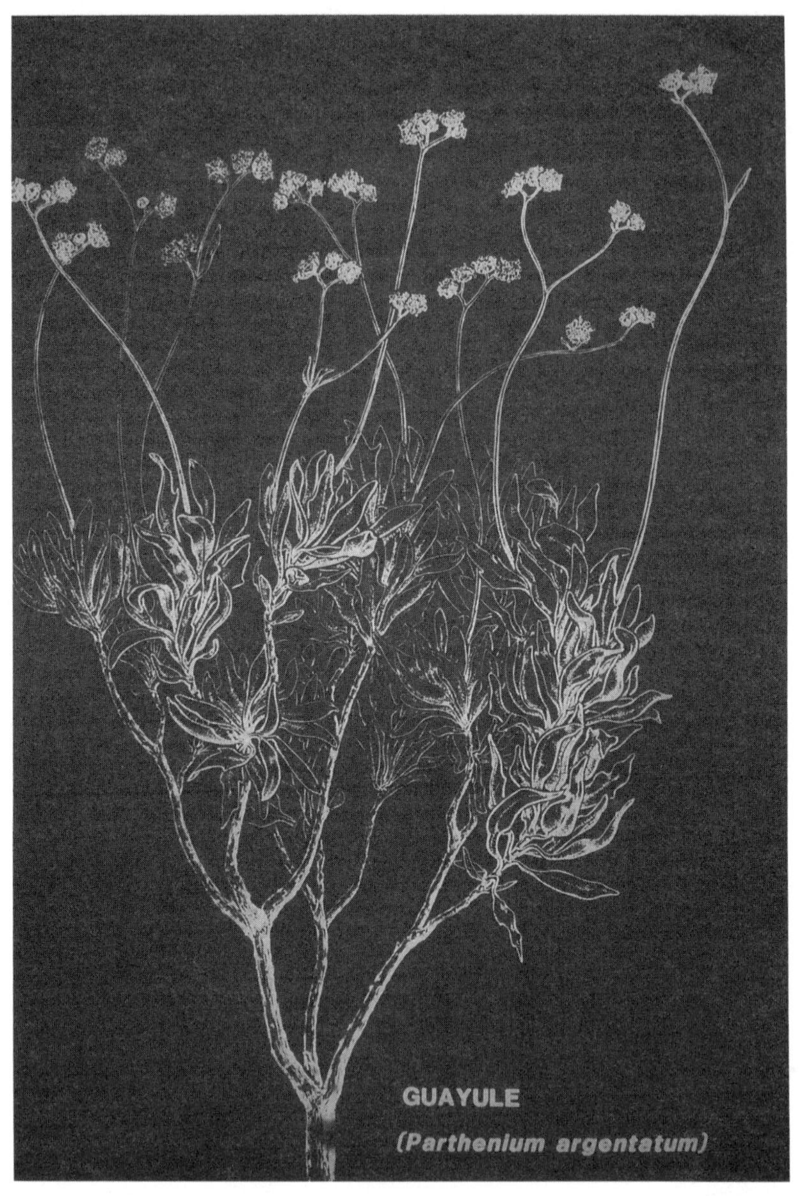

Fig. 5.8. Guayule (*Parthenium argentatum*). Drawing courtesy of Office of Arid Lands Studies, University of Arizona.

usable for traditional crops. It does not require irrigation and can be readily propagated by direct seeding. It can be harvested by mowing and it resprouts from a cut stump, permitting continuous cropping with several harvests through the life of the plant. This characteristic means the plant can be treated as a perennial, thus avoiding the erosion losses associated with annual crops. The strong genetic controls of rabbitbrush suggest that transfer and selection of genes for high rubber and/or resin content can be accomplished rather easily.

Guayule

Guayule (*Parthenium argentatum* Gray) (figure 5.8), usually pronounced wy-oo-lee, is a rubber-producing shrub that is native to the Chihuahuan Desert in southwestern Texas and northern Mexico. Although guayule is a member of the *Compositae* (daisy family), its rubber is a *cis*-polyisoprene type similar to that produced by rabbitbrush and the equatorial rain forest tree (*Hevea brasiliensis,* Euphorbiaceae). Unlike *Hevea,* but like rabbitbrush the rubber in guayule is located in the parenchyma of stems, branches, and roots. It cannot be "tapped" as *Hevea* rubber latex can, but must be extracted from the plant tissues. It contains no natural antioxidant, so the plant must be processed within a few days of harvesting to avoid degradation.

Both guayule and *Hevea* have been known to industry for a long time as sources of natural rubber. Early in the twentieth century guayule rubber was produced commercially from wild stands by both Mexican and American industrialists from fourteen facilities, mostly in Mexico. Maximum production of nearly 11,000 tons was reached in 1910, providing 10 percent of the world's rubber and 50 percent of the U.S. consumption. Figure 5.9 shows a cultivated field of guayule. This production ceased in the 1930s, but during World War II when the *Hevea* rubber supplies from southeast Asia were cut off by the Japanese, there was renewed interest in guayule. Personnel of the USA Emergency Rubber Project planted over 30,000 acres with guayule in Arizona, California, and Texas. In 1946 with the war over and *Hevea* rubber available again, the 77th Congress terminated the project even before most of the experiments could be completed. All 30,000 acres of guayule in the plantations (an estimated 10,000 tons of rubber) was burned or

Fig. 5.9. Guayule (*Parthenium argentatum*). Cultivated field in flower, located near Sacaton, Arizona.

plowed into the ground. Moreover, the wartime synthetic rubber program was quite successful, so most rubber requirements could be met with synthetic elastomers prepared from petroleum feedstocks, or so it was thought.

By the early 1970s it was apparent that many rubber applications were not adequately met by synthetic rubber. This was particularly true for various military and aviation needs as well as for high-performance tires. Some applications require wholly natural rubber, others a blend of natural and synthetic rubbers. The United States now imports all its natural rubber, over 825,000 tons annually, at a cost of about $1 billion. Most of that rubber comes from countries in Southeast Asia. Because of this, and remembering the crisis of World War II, natural rubber is considered a critical and essential raw material.

Interest in the United States in guayule was renewed in 1975 for several reasons: a search for crops that could be grown with less water had been undertaken; natural rubber had gained the status of a strategic material; and fears of dependency as a result of the oil embargo earlier in the decade. The latter increased petroleum prices dramatically, which in turn resulted in an increase in the price of synthetic rubber. In 1978, the U.S. Congress passed the Native Latex Commercialization and Development Act, whereupon work was initiated by a number of state and federal agencies, and the first real steps were taken for possible establishment of a U.S. natural rubber industry based on guayule.

To be more specific, in addition to continuing research efforts at several universities, the Gila River Indian Community, located near Phoenix, Arizona, obtained a contract from the Department of Defense to grow guayule and produce rubber from it. The Indians awarded a subcontract to the Firestone Tire and Rubber Company to design and build a prototype guayule rubber-processing plant on tribal lands (figure 5.10). Firestone conducted pilot plant operations in its facilities in Akron, Ohio and built a prototype production facility at Sacaton, Arizona, near the site where the guayule for the plant is grown by the Indians. The Indians committed to produce enough guayule shrub to produce one thousand tons of rubber annually. The prototype production facility was dedicated in January 1988 and was designed to produce 2,000 pounds of rubber an hour at full production. It was expected to operate long enough to produce testing quantities of rubber and generate economic data to see if guayule could be competitive. The results to date are inconclusive, but encouraging.

Guayule is a semidesert shrub. It can be killed by freezing and is not very salt-tolerant, dual circumstances that limit the areas in which it can be grown. It is a bushy perennial usually about 3–6 feet in height and having narrow leaves, clusters of small inflorescences on long stems, and a root system that is adequate in depth. Guayule rubber is contained in a single layer of thin-walled cells in the stems and in the roots, and may accumulate for at least ten years, with the plants able to retain this rubber for even longer periods.

Guayule produces rubber only when the plants are under stress, although recent studies indicate that the use of bioregulators may alleviate this shortcoming. Irrigation must be employed to establish adequate stands and to develop plants, and then soil

Fig. 5.10. Prototype guayule processing/extraction plant operated by the Bridgestone/Firestone Company near the Gila River Indian Reservation where the guayule is grown.

water levels must be controlled to limit vegetative growth and enhance rubber production. Seeds are harvested mechanically (see figure 5.11) from mature plants. Seedlings are grown in a greenhouse (figure 5.12) and transplanted using mechanized equipment (figures 5.13 and 5.14). Reliable data are not yet available on water requirements or on plant stress effects on rubber production rates.

Weed control in nurseries was one of the most expensive operations in early efforts to grow guayule, so selective chemicals and related weed control techniques are now being developed. Disease problems occur mainly on poorly drained soils or when well-drained soils are over-irrigated. Most diseases of seedlings are caused by fungi, but disease-resistant varieties can probably be developed, as has been done with other crops. Similarly, biological and chemical means of control of insects and other pests are being developed.

Top: Fig. 5.11. Guayule (*Parthenium argentatum*) seed harvester. This four-row harvester uses a vacuum to collect the tiny seeds, while chains help to loosen seeds.
Bottom: Fig. 5.12. Guayule (*Parthenium argentatum*) plantlets are grown in the greenhouse. In photo these are being prepared for the planter. Note mature guayule field in the background. Photo by George Abel.

Top: Fig. 5.13. Guayule (*Parthenium argentatum*) seedling planter. A person removes single plantlets from container on rack and places them into each of the four cones. Machine then automatically places the seedling in the furrow and covers its roots with soil. Photo by George Abel.
Bottom: Fig. 5.14. Guayule (*Parthenium argentatum*) seeding planter in operation showing plantlets in the ground. Note racks holding flats of plantlets which are placed in cones by operator. Photo by George Abel.

Fig. 5.15. Guayule (*Parthenium argentatum*) is harvested by first cutting the plant either several inches above the ground for coppicing or below surface as a final harvest to retrieve roots as well as stems and foliage. The biomass is then windrowed and baled. Photo shows bale being loaded for transport to processing plant. Photo by George Abel.

Harvesting, storage, and processing practices have been modernized to utilize modern machinery (see figure 5.15) in the production of rubber from both stem and root of guayule. It has been shown that rubber yields are highest when the shrub is clipped about 2–4 inches above ground and harvested, and then sacrificed completely (including part of the roots) from one to four years later. Storage practices need to be optimized, too, as both rubber quality and extractable yield appear to decrease with storage. Processing techniques have been studied at length.

Because guayule rubber is associated with resins and other potentially valuable products, any commercial process must be capable of producing each in a relatively pure state, although to do so is an economically demanding task. To illustrate the point, the early commercial method of obtaining guayule rubber utilized a water flotation technique. In flotation processing, ground shrub is agitated in a dilute caustic solution. Resinous rubber "worms" form and are skimmed from the surface. The worms are washed with a polar organic solvent to remove the resin. This system is inherently cumbersome, inefficient, and requires the disposal of large volumes of caustic flotation medium, which is environmentally unacceptable. The next process employed sequential solvent extraction. In this system the shrub is first extracted with a deresinating solvent and then extracted with a second solvent to remove the rubber. A significant benefit of sequential extraction is the improved quality of the resin-free rubber. The present process of choice is one employing extraction of the shrub with two or more solvents simultaneously. In this simultaneous extraction process the ground shrub is extracted with a solvent in which both rubber and resin are soluble, followed by the addition of a polar solvent which coagulates the high molecular weight rubber polymer. Sequential additions or removal of solvent(s) permit separation and isolation of relatively pure fractions of coproducts. This technique also provides better solvent recovery than sequential extraction. More detail on these procedures is provided in chapter 6.

The single most important deterrent to guayule commercialization has been, and continues to be, rubber yield. Much has been accomplished during the past few years and there is now good reason to believe that high-yielding guayule strains can be developed for a variety of geographic locations. Present yields of rubber and resins vary widely for the same strain grown in different locations. Growth habit and biomass also differ greatly from strain to strain and from accession to accession. Fortunately, the germ plasm pool is large and varied. Some skilled plant breeders (figure 5.16) feel it is highly likely that the present average yields of about 400 pounds of rubber per acre per year can be at least doubled. Then, too, most guayule varieties produce nearly as much resin as rubber. Recent work indicates the resin may have significant economic value, thus helping to offset the cost of the rubber.

Before one can determine the commercial viability of guayule a thorough knowledge of all significant potential products and the economic value of each must be in hand. Guayule produces both high and low molecular weight rubber, resin (complex mixtures), and bagasse. Resin composition varies with shrub line, cultivation

Fig. 5.16. Mr. George Abel, project manager of the Gila River Indian guayule development program, stands between two different guayule cultivars, both of which are ready to be harvested.

site, harvest date, and processing history. The high molecular weight rubber is used in traditional applications just like *Hevea* rubber. The low molecular weight rubber is useful as a plasticizer or processing aid, while the low viscosity coproduct polymer can serve as feedstock for production of depolymerized rubber, which is widely used in adhesive and molded-product manufacture.

Guayule resin shows promise as a wood protectant in both marine and terrestrial environments. It can also be used as a plasticizer for high polymers. The resin seems to hold potential for utilization as a chemical process feedstock for coatings, rubber compound additives, and catalytic conversion to fuels or secondary feedstocks.

The bagasse has been considered as a cogeneration fuel, a feedstock for gasification and conversion to liquid hydrocarbons, and a source of fermentable sugars or fiber. The latter two are not good candidates, however. The most likely use for guayule bagasse at present is as a cogeneration fuel.

In summary, progress has been made on all guayule fronts. Although to date a sufficiently high rubber-yielding strain or accession has not been demonstrated, the prospects of developing one appear promising. Much progress has been made in all aspects of guayule agronomy, harvesting, and processing. Although it is doubtful guayule is economically viable today without some sort of subsidy, if research continues it may become a new crop within a few years.

Milkweeds

The common *milkweeds* (*Asclepias* species) produce latex-containing chemicals that offer potential as sources of rubber, chemical feedstocks, and fuel. The residue after extraction is rich in nutritious and digestible protein. There are literature reports as early as 1875 on the manufacture of rubber from milkweed. At least a dozen *Asclepias* species that grow wild in North America have been reported to produce hydrocarbons including rubber.

Scientists at the United States Acclimatization Garden near Bard, California conducted investigations with *Asclepias erosa* from 1931 to 1934, and reported that the leaves of wild plants collected in Yuma County, Arizona contained from 2.45 to 13.06 percent rubber. In the 1970s the plant was examined as a potential hydrocarbon fuel source. *Asclepias erosa* is a herbaceous perennial averaging four feet or more in height when mature. New stems are produced in March, grow rapidly through spring and early summer reaching maturity and maximum rubber content by fall. The dried leaves of this species can be stored in a dry room under normal light conditions for 18 to 20 months without appreciable loss in rubber or hydrocarbon content. Seeds collected from the wild plants germinate readily when planted in the field. Seedlings are grown in pots and successfully transplanted, but the increase in germination is not great enough to justify the time and labor involved. Preliminary experiments indicate that selection of high rubber-yielding strains is feasible.

In the late 1970s and early 1980s, studies were conducted on *Asclepias speciosa* at the Plant Resources Institute in Salt Lake City,

Fig. 5.17. A planted field of milkweed (*Asclepias speciosa*) in full bloom, located near Syracuse, Utah. Picture courtesy of Native Plants, Inc., Salt Lake City, Utah.

Utah. These studies included evaluation of agronomic problems of seedling establishment, weed control, harvesting, and storage of material until processing. Chemical analyses revealed the presence of triterpenoids and *cis*-1,4-polyisoprene rubber in the hexane extract and polyphenols, inositol, and sucrose in the methanol extract. The residue after extraction was found to be high (about 16 percent) in good-quality digestible protein suitable for ruminating animals. The amino acid composition of a sample of the protein from a late June harvest was comparable to that of alfalfa and generally superior to that of corn grain. Irrigation did not appear to be important except to get the crop started. In one experiment, the yield of hexane extractables was actually 74 percent larger in the dryland vs. the irrigated plots. A cultivated field of milkweed is shown in figure 5.17. For harvesting, conventional haying practices were successful. The milkweeds were cut, crimped, and windrowed in one operation, then field-dried for three or four days and baled with conventional equipment. Alternatively, the milkweeds were green-chopped with a silage cutter and ensilaged. Storage tests on the bales revealed no significant differences in yields of either extract after two months of outside storage at ambient Salt Lake City temperatures. The milkweed silage was stable after three months, although there was a shift of methanol extractables to the less polar hexane extractables.

Asclepias species produce large seed pods containing many seeds with a pappus of silky hairs attached at the hilum. This silky down permits the seeds to be scattered by the wind and represents another item of potential commercial importance. The Natural Fibers Corporation of Bay Village, Ohio currently has over 200 acres of milkweed under cultivation in Ogallala, Nebraska. Photos of the blossom and fibers are shown in figures 5.18 and 5.19. They harvest the milkweed pods and process them to recover the white seed hairs or floss. Their near-term market is a mixture containing 30 percent milkweed floss with 70 percent goose down. This floss-enhanced down, called "Premium Ogallala Down," is marketed as a loose-fill insulation for comforters, jackets, and sleeping bags. Another product for the intermediate term is the formation of milkweed floss into a nonwoven fabric 60 inches in width and about ¼ inch thick. They have made 4,000 linear yards of this batt material for test marketing. This product is lighter and warmer per unit of thickness than Thinsulate™.

Milkweed floss is a very fine fiber, but it is not dusty and is easily retained in regular down-proof fabrics. It appears to be non-allergenic, allows perspiration to evaporate at a 50 percent higher rate than goose down and can be cleaned by either washing or dry cleaning. Longer term, with increased production and lower costs, Natural Fibers Corporation considers the possibility of milkweed floss competing with pulp in paper products. The milkweed fiber is very soft and can absorb up to 80 times its own weight in liquid.

Everything considered, the milkweeds offer real promise as multi-use energy, chemical, fiber, and feed crops for the semiarid regions of the world. They would seem to be particularly appropriate as an alternative native crop for the Great Plains overlying the Ogallala aquifer.

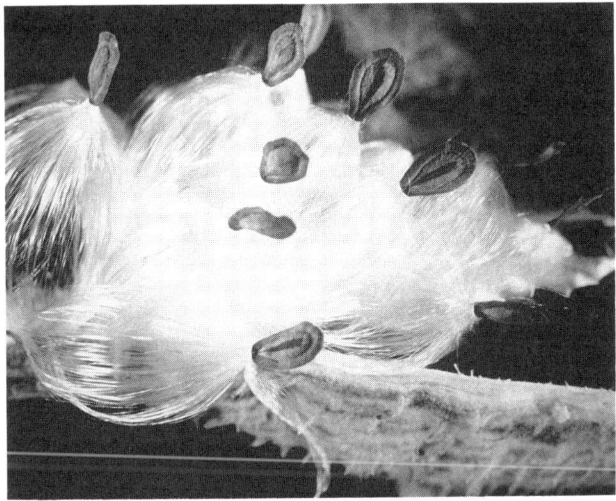

Top: Fig. 5.18. Milkweed (*Asclepias speciosa*) flower. Courtesy of Native Plants, Inc., Salt Lake City, Utah.
Bottom: Fig. 5.19. Note the pappus (fiberous filaments) and connecting seeds of the milkweed (*Asclepias* spp.). The fibers are used commercially as an insulating floss. Picture courtesy of Dr. Jess Martineau.

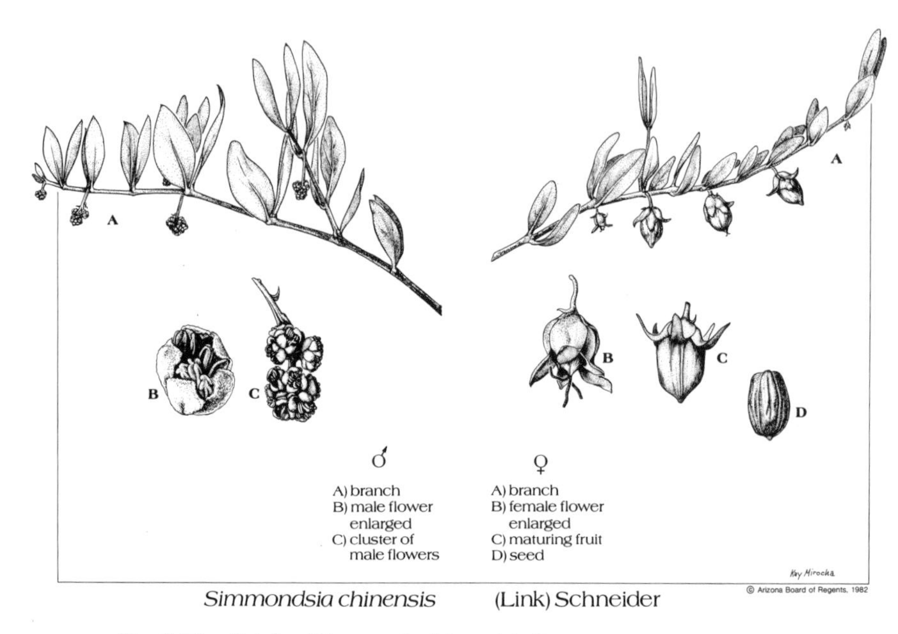

Fig. 5.20. Jojoba (*Simmonsia chinensis*). Drawing courtesy of Kay Mirocha, Office of Arid Lands Studies, University of Arizona.

Jojoba

Jojoba (pronounced ho-ho-ba) (*Simmondsia chinensis*) (figure 5.20 is a hardy evergreen shrub native to the arid regions of northern Mexico and the southwestern U.S. The species is heat-tolerant, capable of withstanding ambient air temperatures of 120 degrees F. and soil temperatures of 150 degrees F., but with limited tolerance to frost. Mature plants can survive 16 degrees F. at least for short periods of time, but flower buds and seedlings can be damaged at 28 degrees F. and killed at 22 degrees F. They survive severe water stress and many strains of jojoba are tolerant of salinity, but the influence of these factors on seed yield require careful study.

Jojoba seeds contain a liquid wax often referred to as an oil because it is a liquid at ambient temperatures. Chemically the oil is a wax ester consisting mainly of a long chain (20 carbon atoms) fatty acid and a long chain (22 carbon atoms) fatty alcohol. Both the fatty acid and the alcohol are monounsaturated. This oil is unique in that it differs from typical vegetable oils, which are mainly triglycerides, i.e., three fatty acids esterified onto a glycerol backbone. Jojoba oil is stable under conditions of extreme heat and pressure, making it a superior lubricant. Neither does it become rancid under extreme conditions like other vegetable oils. With these properties it is an ideal replacement for sperm whale oil, which was so valuable to industry at one time that the U.S. classified it as a strategic material and stockpiled it against a national emergency. Today the sperm whale is protected as an endangered species and all whale products are banned from importation. Jojoba oil has been shown to be not only an ideal replacement for sperm whale oil, but also superior for many applications. However, very little jojoba oil has ever been sold as a replacement for sperm whale oil because industry discovered cheap although inferior replacements for it.

The outstanding properties of jojoba oil make it a highly desirable candidate for many applications including cosmetics, pharmaceuticals, lubricants, and transmission fluids, but the supplies have been limited and unpredictable, and the prices are too high for many potential uses. Wild jojoba stands provided the first commercial quantities of seeds for research and product and market development. It has been estimated that the wild plants in the U.S. and Mexico produce between 22,000 and 35,000 tons of seed per year, but only a small amount is actually harvested because of the inaccessibility of the plants, the high cost of harvesting, and land ownership problems. Also, the seed yields of wild plants fluctuate widely from year to year with changes in the local weather conditions. Only plantations hold the potential of producing jojoba inexpensively enough to attract large volume uses.

It was not until the 1970s that the broad industrial and agronomic potential of jojoba began to attract public attention. In 1971 the first significant harvests were reaped from wild stands, mainly by Indians collecting and processing seed from their reservations in Arizona and California. Basic research results on the crop stimulated farmers, environmentalists, and entrepreneurs to undertake the serious task of converting the wild shrub to a commercial crop. Several universities, corporations, government, and private research laboratories in the U.S., Mexico, Israel, Australia, and other countries began efforts to determine the basic agronomic requirements for jojoba.

Jojoba plants are usually either male or female, and only the female plants produce seed (figure 5.21 shows female jojoba flow-

Fig. 5.21. Clustering variety of female jojoba flowers. Photo courtesy of Jojoba Growers & Processors, Inc.

ers). It was necessary to determine the best ratio of male to female plants to ensure proper fertilization and optimum seed yield. Optimal amounts of water and fertilization had to be studied as well as efficient methods of harvesting and control of pests and diseases. Plantations were established using direct seeding, seedlings, rooted cuttings, and plantlets produced from tissue culture. Wild jojoba is genetically extremely heterogeneous, so a major effort was required to select uniform high-yielding varieties suitable for monoculture plantations and mechanical harvesting. A favorite cultivar is shown in figure 5.22. The plants must be three to four years old before they bear seeds and about ten years old for maximum seed production. This imposes a long-term investment before any income can be realized. Fortunately, the plants are believed to be long-lived: 100 to 200 years. Examples of mechanical harvesting equipment are shown in figures 5.23 and 5.24.

Fig. 5.22. This jojoba (*Simmonsia chinensis*) cultivar named "Keiko" for Keiko Purcell, shown in picture, is reputed by its developer, to be the best cultivar now available. Its upright growth habit provides easily harvested high yields of high oil-content fruit. Note the elevated drip-irrigation system. Photo by the Purcell Jojoba Company.

Fig. 5.23. Jojoba (*Simmonsia chinensis*) has been harvested by several different mechanical means in addition to hand picking. Shown here is the first commercial "from the ground" mechanical harvesting system. Photo provided by Purcell Jojoba Company.

By 1989 significant progress had been made toward solving most of these problems. In the U.S., just ten years after the first major commercial plantings were made, the 1988 harvest totalled nearly 3 million pounds. The average yield was only 177 pounds per acre, but many growers reported yields of 400–500 pounds per

Fig. 5.24. Jojoba (*Simmonsia chinensis*) harvesting machine showing the seed/rock separator tumbler and the hand-operated vacuum pick-up hoses. Photo provided by Jojoba Growers & Processors, Inc.

acre, and one 9-acre test plot provided over 2,000 pounds per acre. In 1988, 13,000 acres were harvested in Arizona and California, compared with only 7,000 in 1987. The total U.S. acreage in jojoba plantations has been as high as 40,000, but is now much lower. Many have been abandoned either because of poor management or inappropriate agronomic practices including starting the plantation with unsuitable, poor-yielding plants. Frost too has been devastating to some inappropriately located plantations. Some of the better farmers will survive and expand their plantations with proven seed stock and good agronomic practices. In addition to the U.S. plantations, more than 30 other countries are raising jojoba. Mexico, Costa Rica, Brazil, Argentina, Paraguay, and Australia each have more than 5,000 acres in jojoba plantations. Israel and China each have 1,000 acres or more and a number of other

Fig. 5.25. Jojoba oil processing plant showing presses, oil drums, and seed storage bins. Photo provided by Jojoba Growers & Processors, Inc.

countries have smaller plantations, bringing the worldwide acreage to between 70,000 and 75,000. When all of this acreage matures to full productivity within a few years, the supply of jojoba materials should be dependable and sufficient to support an impressive industrial development. Figure 5.25 is a photo of a jojoba oil processing plant.

In 1981 the United Nations Industrial Development Organization predicted the total world annual demand for jojoba oil would reach 65,000 tons by 1995 when the price of the oil was expected to decline to about $4,000 per ton. Today, experts say there is no good reason to modify this demand forecast, but there will probably be a ten-year delay in achieving the production volumes required to meet the predicted demand. The present price for jojoba oil is almost twice the projected price. The delay has been attributed to a number of factors. In the U.S. it was a combination of underestimating planting and cultural costs, overestimating seed yields and market prices, and unusual weather conditions in 1986 and 1987. Internationally, the delay was caused by later than predicted planting in Latin America and other low-cost producing countries. Also, there was less than optimum research and development for defining value-added jojoba derivatives for industrial uses.

Until now, over 90 percent of all jojoba oil on the world market has been used in the cosmetics industry in hair and skin preparations as a 1 to 3 percent emollient. Jojoba oil began to be used in cosmetics products in the mid-1970s by a few small companies, but now most major cosmetics manufacturers offer at least one product containing jojoba. In the U.S. alone, over 300 products containing jojoba oil are available, but many of them contain only a trace amount to take advantage of the "fad" value. In addition to the mystique associated with the name and its attraction as a "natural" ingredient, jojoba oil helps make skin smoother and softer, penetrates better than most oils, is nongreasy, nonirritating, odorless, and is a good moisture barrier with excellent stability. Its surfactant properties make jojoba oil especially desirable in shampoos because it makes hair feel thicker and less tangled, and it reduces the incidence of split ends. As these fundamental properties become better known and appreciated, and with increasing production and lower prices, it is estimated that the cosmetics industry will use between 10,000 and 12,000 tons per year. Examples of jojoba cosmetics products are shown in figure 5.26.

Jojoba oil and its derivatives have outstanding potential as a lubricant additive for use in automobiles, trucks, buses, compressors, magnetic media, jet engines, defense vehicles, and other heavy machinery. Jojoba-based engine lubricants reduce friction between moving parts and can reduce wear by as much as 43 percent. The addition of as little as 3 percent jojoba oil to transmission fluids can reduce operating temperatures by as much as 30 to 50 degrees F. Addition of 5 percent jojoba oil to automatic transmission fluid gives a smooth sensation and provides a very positive effect upon wear reduction. Sulfurized jojoba and a polar derivative are useful as additives to hydraulic fluids, lubricants, and transmission products. It can also be used to extract mercury from aqueous solutions and should be an extractant for other metals such as copper and zinc.

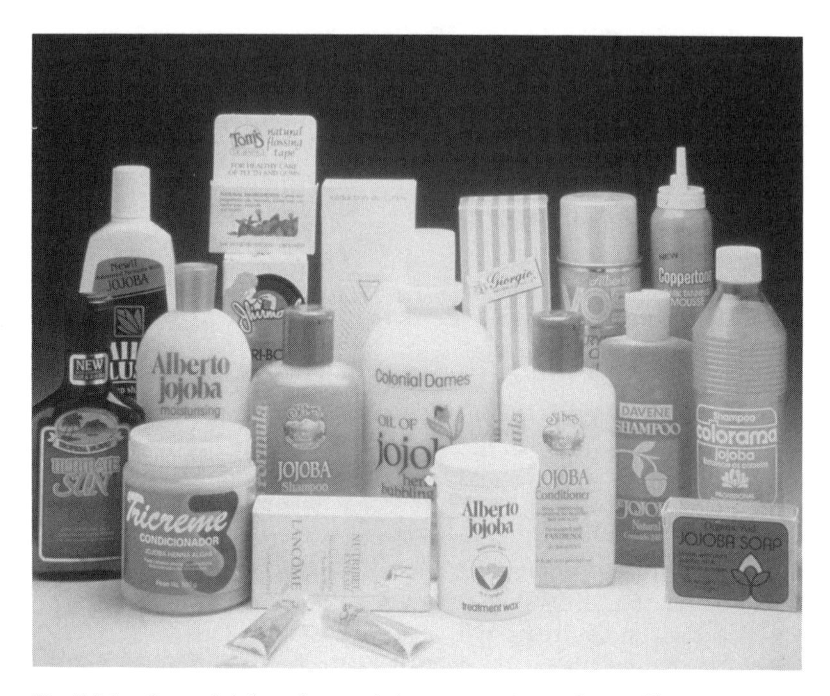

Fig. 5.26. Some jojoba oil-containing cosmetic products. Photo provided by Jojoba Growers & Processors, Inc.

Small-scale commercialization has been initiated and customer acceptance indicates that the jojoba-containing products are significantly better than the traditional products available. Bob Anderson of International Lubricants, Inc. reported that his company used over 100 fifty-five-gallon drums of jojoba oil during 1988 and they expected to need from 250 to 300 drums during 1989. Based on projected sales figures, the requirement by 1993 will be between 300,000 and 400,000 gallons. Jojoba-fortified lubricants meet the demanding requirements of the popular front wheel drive, turbocharged automobiles better than any other readily available material. When the major oil companies introduce widespread use of jojoba oil in automotive lubricants, it could have the very beneficial effect of reducing U.S. petroleum consumption by millions of barrels each year.

Jojoba oil offers attractive possibilities to the pharmaceutical industry. Its high purity, lack of odor, and resistance to rancidity make it an excellent base for ointments and creams. When supplies increase and prices drop, it could well displace mineral oil and lanolin for these uses. Jojoba oil is an excellent antifoam agent, offering an interesting possibility in the manufacture of antibiotics. Comparative tests have shown jojoba oil to be a superior antifoam agent at least for production of penicillin G and cephalosporin. When supplies and prices permit, this specialized market could absorb 8,000 tons of jojoba oil annually in the U.S. alone. Jojoba's potential for direct use as a pharmaceutical is being investigated as a treatment for burns, acne, and psoriasis. Preliminary results appear favorable, but these applications require extensive clinical study and approval by the U.S. Food and Drug Administration and similar international agencies before a product can be marketed.

Jojoba and its derivatives have potential in other applications, including use in transformer oils, electrical insulators, disinfectants, detergents, wax for candles, and polish for cars, floors, furniture, and shoes. It may serve as an acceptable replacement for sperm whale oil in the leather industry as a softening and tanning agent. In the textile industry it shows promise as a lubricant in the manufacture of synthetic fibers by reducing fiber-to-metal friction. Jojoba oil is an excellent surfactant, i.e., it reduces the surface tension of a liquid in which it is dissolved. Many consumer products require surfactants. Kelley Dwyer, vice-president of Jojoba Growers and Processors, Inc. of Apache Junction, Arizona has predicted that by 1992 this use will require 400 to 600 barrels annually.

The properties of jojoba oil make it a likely candidate as a cooking oil. It is not absorbed by food as much as most cooking oils. It is bland tasting, heat-stable, does not become rancid, and is resistant to hydrolysis by digestive enzymes which hydrolyze most fats. This means that jojoba oil is digested much less than conventional cooking oils and accordingly should make a healthful, low-calorie food oil. In addition to use as a cooking oil it can be used directly to prepare salad dressings, etc. For centuries the Indians of the Southwest used jojoba in various forms as a valued food, but before it can be marketed in the U.S. it will require FDA approval. The Jojoba Marketing Cooperative has entered a petition stage with the FDA to get guidelines for obtaining this approval.

The meal left after jojoba oil has been expressed or extracted is a potentially valuable by-product. It contains about 25 percent protein as well as carbohydrate and fiber. The essential amino acid content of the protein is generally good, being high in lysine, but is deficient in methionine.

The main problem with the meal as an animal feed is that it contains four compounds known as simmondsins, which make it unpalatable and potentially hazardous. Fermentation using the acidophilous bacterium of sweet milk or ensilage offer two methods to detoxify the meal and make it a palatable, nutritious feed. But, before the meal can be marketed in the U.S., FDA requirements to show that the hazardous compounds cannot be transmitted to milk, meat, or eggs will have to be met. The fiber from jojoba meal has gained some commercial success in two applications. A fine grade of the fiber is being used by a few cosmetic manufacturers as a mild abrasive in facial scrub products, and a coarser grade is being sold for use in radiator "stopleak" products.

In the U.S., 1988 was a banner year for jojoba. In an interview printed in the January-February issue of *Jojoba Happenings*, Vicki Hubbard, the president of the Jojoba Growers Association (JGA), said, "I think for the first time in a decade, growers and processors can look back and generally breathe a sigh of relief. . . . We've seen the production of at least two million pounds of clean seed. Our markets have increased substantially. Processors are paying good prices for the seed and they are marketing the oil in a highly professional manner." During 1988 frost damage was minimal, a favorable tax ruling was obtained from the Internal Revenue Service, and the JGA succeeded in registering three badly-needed pesticides. With success in the plantations and processing, further development of jojoba now depends more on the creativity of the chemical industry and its ability to utilize the unique natural products now available for the first time from jojoba. The shrub grows in soils of marginal fertility, requires less water than most crops, is tolerant of desert heat and soil salinity, and provides the world with a new, renewable resource that can fill many of humanity's needs.

Buffalo Gourd

The virtues of this plant (*Cucurbita foetidissima*) as a food crop were described above in chapter 4. *Buffalo gourd* also qualifies as an arid land candidate for the production of chemicals, alcohol, and fuel. As pointed out earlier, this species with little or no irrigation produces large yields of high-starch-content roots and high-oil-content seeds.

There are two cultural systems for buffalo gourd production.

The perennial mode optimizes seed yield with limited root yield, while the annual mode optimizes root production with little seed yield. In the perennial mode initial plantings of 600–2,400 plants per acre provide for seed harvest annually, and one-half of the roots would be harvested by digging alternate yard-wide swaths in order to thin the plantation and prevent overcrowding. During the growing season the harvested swaths regenerate by asexual rooting of the vines creeping over from the unharvested swath. Figures 5.27 and 5.28 illustrate the difference in size achieved by roots after a single growing season vs. three years of growth.

The seed oil, which is very similar to soybean oil, offers some potential as a diesel fuel extender and as a chemical raw material. The triglycerides of the seed oil contain about 27 percent oleic acid. This compound can be split chemically into pelargonic acid and azelaic acid which are used in many industrial applications including plasticizers, lubricants, hydraulic fluids, specialty polymers, and heat transfer fluids. Soybean oil is already an important ingredient in protective coatings. Buffalo gourd seed oil is similar in composition and has been shown to compare favorably for use in paints and varnishes. The protein from the gourd seed has been evaluated and several potential uses for it are indicated: water-based paints, paper coatings, adhesives, and textile sizings. And, the root starch, in addition to serving as a food and as the substrate for the production of ethanol, has been found to offer possibilities as a component of biodegradable plastics, adhesives, and sizings.

For root-yield optimization, buffalo gourd is planted at a much higher density (200,000–250,000 plants/acre) in the spring, and is harvested in the fall. Root yields of over 30,000 lbs./acre have been achieved without any agronomic optimization. From these an average root-starch content of 63.5 percent (dry-weight basis) was obtained. The moisture content of freshly harvested roots is 67–69 percent. Even though roots harvested annually are much smaller than those harvested after two growing seasons, root yields per acre are substantially larger. Additionally, smaller roots are easier to harvest and process than larger ones and thus are more economically attractive. The roots so obtained can be utilized as indicated above, or burned as a good and abundant cooking or heating fuel. Buffalo gourd starch as an edible product has been discussed previously (see chapter 4).

If the roots are to be used to produce alcohol they must be washed, chopped, dried, and finely ground. This series of operations could be done at the fermentation site or apart from it as

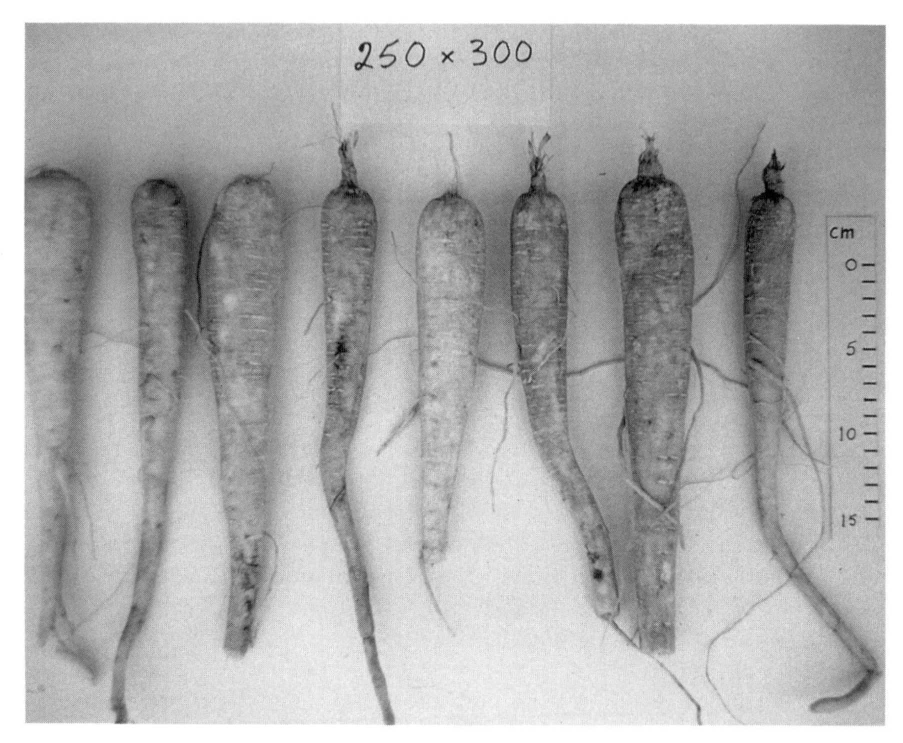

Fig. 5.27. Buffalo gourd (*Cucurbita foetidissima*) roots harvested after only one season of growth. Roots of this size are best for replacement of wood in cooking and heating (rootfuel). Photo by Dr. Eugene B. Shultz, Jr.

expedient. In any case the root "flour" is saccharified and subsequently fermented to produce ethyl alcohol.

Using the root starch yield reported above, alcohol yields may be estimated at 400 gallons/acre, based on the standard industry conversion of 15 pounds of starch to one gallon of alcohol. If this is verified in practice, buffalo gourd would be comparable to sugar beet and better than either corn or grain sorghum as an alcohol feedstock. Furthermore, buffalo gourd requires neither the quality of land nor the quantity of water required by sugar beets, corn and grain sorghum. Icerman and Shultz (1985) have carried out preliminary economic analyses of buffalo gourd as an energy crop in the high plains of eastern New Mexico, an area dependent on the Ogallala aquifer for its irrigation water. In their comparison of buf-

Fig. 5.28. One-half of a three-year-old buffalo gourd root. This specimen weighed 319 pounds. Although this root was produced by a plant growing under extremely arid conditions, the 154 pounds of starch it contained is equivalent to the amount produced by a score of potato plants grown under good conditions. (Photo by L. C. Curtis.)

falo gourd, grain sorghum, corn, and wheat, the buffalo gourd exhibited the lowest cost for producing the amount of crop needed to make one gallon of liquid fuels. Net revenue from buffalo gourd was projected to be substantially higher than that for grain sorghum, the nearest competitor.

Fermentation of starch to produce ethyl alcohol also produces coproducts. The solid residual (dried distillers grain, or DDG) from the fermentation of corn and other grains is sold as an animal feed. The stillage, i.e., the residual liquid from alcohol distillation, contains soluble nitrogen compounds and is disposed of in any of several modes, some leading to groundwater pollution. Because the solid residual (DDG) from buffalo gourd contains high levels of bitter cucurbitacins it cannot be utilized as a feed without further treatment. Any such treatment adds direct cost to the product and capital cost to the facility. It has been suggested by Sl.ultz and Arding (1987) that the entire problem might be avoided by converting all residuals from the production of alcohol to fuel gas. This is accomplished through anaerobic fermentation which yields a mixture of methane and carbon dioxide that can serve as a fuel gas for operation of the plant, explained further in chapter 6. In addition, by use of the Anamet Process (Skogman 1979), nitrogen in the liquid effluent may be removed as ammonium salts. Such a scheme lowers the overall capital cost of the processing plant, permits the marketing of a single major product (ethanol), generates much of the energy required to run the plant, and mitigates groundwater pollution problems that continue to plague the present fuel ethanol industry.

Converting buffalo gourd roots into ethanol as a single product appears to be economically attractive for both the farmer and processor. Ecologically it is attractive too, for it would conserve water as an alternative to irrigation farming and probably conserve topsoil as well. However, the likelihood that buffalo gourd will replace any conventional crop in the United States, regardless of how beneficial it may seem, is slight under the government's present farmer subsidy program. The farmer is guaranteed a fair return, even when growing additional surplus grain, and thus has no incentive to grow a new crop, like buffalo gourd, where he must bear the burden of risk. Then, too, the U.S. government spends only a trivial amount on research and development of new alternative crops compared to that on conventional surplus crops. In the United States, there is little likelihood that buffalo gourd will become a new crop of significance until the federal government recognizes the desirability of change in its policies toward agriculture.

Buffalo gourd may well become a significant crop in Third World countries in the near future, however. One reason is that dried buffalo gourd roots are a good cooking fuel. Reforestation and high-efficiency cookstoves are unable to keep pace with the ever-increasing demand for wood fuel in many Third World arid lands. The situation becomes more critical every day. The need for an alternative to wood for cooking is obvious as reforestation is slow and to date has not been very successful.

In dryland ecosystems buffalo gourd will produce almost twice as much biomass in a summer as mesquite does annually on the same area of land. It is resistant to most diseases and easy to grow and harvest with the simplest of tools. Growing on marginal soil and sloping terrain, it can fit into cropping systems without displacing other crops and because of the abundance of roots and large, stiff leaves may serve to mitigate erosion caused by the infrequent but heavy rains characteristic of arid lands. In the simplest system the roots are dug, cleaned, usually split, then sun dried prior to use as fuel.

Although the heating value of this root fuel is about 10 percent less than that of wood it burns more slowly, thus the transfer of heat to the cooking vessel is more efficient. The same boiled meal cooking task is accomplished using about 70 percent as much root fuel as wood by weight. Acceptability tests have been conducted in rural Senegal, Niger, Zimbabwe, and Mexico with generally favorable results. Figure 5.29 shows a Mazahua Indian woman cooking with buffalo gourd root fuel. None of the cooks found objectionable flavors. In terms of heating value and rate of weight loss during burning, root fuel is more like wood than charcoal. However, the ignitability of root fuel is similar to that of charcoal.

A preliminary economic analysis of growing buffalo gourd for root fuel in competition with wood fuel was conducted in the Deccan Plateau of India. This plateau is a dry, deforested and populous region comprising the bulk of the Indian peninsula. The results of the study indicate that root fuel would compete with subsidized wood fuel without the need for subsidy under many, but not all, scenarios investigated.

The prospects for rapid and inexpensive dissemination of buffalo gourd for root fuel use appear promising because of the inherent simplicity of the root fuel concept and the opportunities for

Fig. 5.29. A Mexican Mazahua Indian woman cooking with buffalo gourd rootfuel on her three-stone fireplace. Her home area is almost completely deforested. Use of rootfuel saves her and other women in the area many hours of scavenging for trash wood each day. Photo by Dr. Eugene B. Shultz, Jr.

local control that root fuel provides. It seems likely that local people can grow their own, and local small businesspeople can bring root fuel to local marketplaces, at low cost, all with a minimum of intervention from governments.

Vernonia Species

Epoxy fatty acids are consumed by U.S. industry in the manufacture of plastic formulations, protective coatings, and other products at the rate of more than 70,000 tons per year. European and Asian industry also use epoxy fatty acids. At the present time epoxy compounds are obtained from petrochemicals or are prepared from linseed oil or soybean oil by processes which are expensive in both time and money. During the late 1950s the USDA Agricultural Research Service initiated plant screening programs to identify new sources of industrial raw materials. One program focused on finding unusual seed oils for which new industrial markets might be created. Seeds were screened for high oil content and those with substantial amounts (greater than 20 percent) were evaluated to identify oils of unique fatty acid composition.

By 1960 *Vernonia anthelmintica* (L.) Willd., a plant obtained from the Indian Agricultural Research Institute, was found to have an oil content of 26.5 percent, and 67 percent of the oil was an epoxy oleic acid. This epoxy acid was identified as vernolic acid, *cis*-12,13-epoxy-*cis*-9-octadecenoic acid. The triglyceride of vernolic acid is called trivernolin. The epoxy groups of such triglyceride oils make these materials useful in plastics and coating products. A program was mounted to evaluate *V. anthelmintica* germ plasm with the objective of developing varieties suited to American agriculture. A parallel program was initiated to conduct utilization research on the oil and its components. The esters of vernolic acid were found to compare favorably with epoxy-containing plasticizers used commercially, and refined epoxidized oil and trivernolin were shown to have potential value as primary plasticizers for polyvinylchloride. However, utilization research was discontinued when it became clear that varieties of *V. anthelmintica* suitable for growing in the U.S. were not likely to be developed.

Interest in epoxy acid-producing plants was renewed when Dr. Robert Perdue of the USDA recognized that *V. galamensis* in wild stands in Ethiopia showed no signs of seed shatter, disease, or insect infestation and contained even higher oil and epoxy acid content than *V. anthelmintica*. Private workers carried out agronomic studies in Kenya and Puerto Rico, and several hundred pounds of seed were made available for processing and utilization studies. The oil was shown to produce high-quality baked films and coatings on steel panels. The physical properties of the films were outstanding in that they were flexible, resistant to chipping, could be drilled, cut, or trimmed without loss of adhesion, and had good resistance to alkali, acid, and solvents. Other possible applications include plastics and adhesives. The Environmental Protection Agency's new restrictions on the use of volatile organic solvents enhances the market prospects for vernonia oil because it can serve as a solvent that will not evaporate and pollute the air, but become a part of the product.

Top: Fig. 5.30. At the Chiredzi Research Station in Zimbabwe, Africa, ARS botanist Robert Purdue (left) and agronomist C. T. Nyati inspect a field of *Vernonia galemensis*. Photo courtesy of Dr. Perdue.
Bottom: Fig. 5.31. Agronomist C. T. Nyati of the Ministry of Lands, Agriculture and Rural Resettlement, Zimbabwe, Africa inspects a field of *Vernonia galemensis*. Picture courtesy of Dr. Robert E. Perdue, Jr.

Recently, an agronomic program in Zimbabwe has been very successful. By 1988 oil yields of 928 pounds per acre were achieved using unimproved germ plasm from Ethiopia. Agronomists in Zimbabwe (figure 5.30) believed they could double the yield by better management of germ plasm and possibly triple the yield by breeding better varieties when more germ plasm becomes available. A useful management technique called "topping" was employed in Zimbabwe. The plants are cut off about six inches above the ground, causing the stems to produce 18 to 20 branches from the base. Each branch produces three to five flower heads and they all mature with seed production at the same time. Plant height is reduced and the time from planting to harvest is shortened (figure 5.31). Vernonia oil contains 80 percent trivernolin. In early 1990 yields of 2,227 pounds/acre of vernonia seed were reported compared with 1,926 pounds/acre for soybeans. Great progress continues to be made.

Vernonia galamensis (figures 5.32 and 5.33) may well become a commercial crop for semiarid tropical regions. Robert Perdue reported that this species is likely to grow well in any area that gets 20 to 40 inches of annual rainfall with a wet season 3 to 4 months long, well-drained soil, and a location within 20 to 30 degrees of the equator. This means that varieties currently available would not do well in the continental U.S., but potential areas include Mexico, Brazil, Central America, islands of the Caribbean Sea, Australia, India, and its native Africa.

Bladderpod

Hydroxy fatty acids are important industrial chemicals used in making many products, including plastics and coatings. Castor oil, which contains ricinoleic acid (12-hydroxy-9-*cis*-octadecenoic acid), is the main source of industrial hydroxy fatty acids. The U.S. alone imports some 60,000 to 75,000 tons annually. Castor beans (from *Ricinus communis*) could be raised in the U.S., but the plant has been avoided because of the danger of seed toxicity, allergic reactions of field workers, and the liability of disposing of the toxic meal left after extraction of the oil. A screening program at the Northern Regional Research Center in Peoria, Illinois identified many plant species that could be used to produce hydroxy fatty acids. Some of these plants produce ricinoleic acid, but not at the high level found in the castor plant. Other plants were found that

Top: Fig. 5.32. Flower head of *Vernonia galemensis.* Photo by Dr. Robert E. Purdue, Jr.
Bottom: Figure 5.33. Seeds of *Vernonia galemensis.* Photo by Dr. Purdue.

Fig. 5.34. Bladderpod (*Lesquerella fendleri*). Reprinted with permission of *Scientific American.*

produce substantial quantities of other hydroxy fatty acids which could function as substitutes for ricinoleic acid or have certain advantages over it.

Members of the genus *Lesquerella* in the Cruciferae family were found to produce high levels of hydroxy fatty acids in their seed oils. Several members, including *Lesquerella fendleri* (figure 5.34) (common name: bladderpod or popweed) offer potential for becoming a commercial crop. The seeds of this bladderpod contain 20 to 30 percent oil and the oil contains about 60 percent lesquerolic acid, a 20-carbon monohydroxy monoene fatty acid that is very

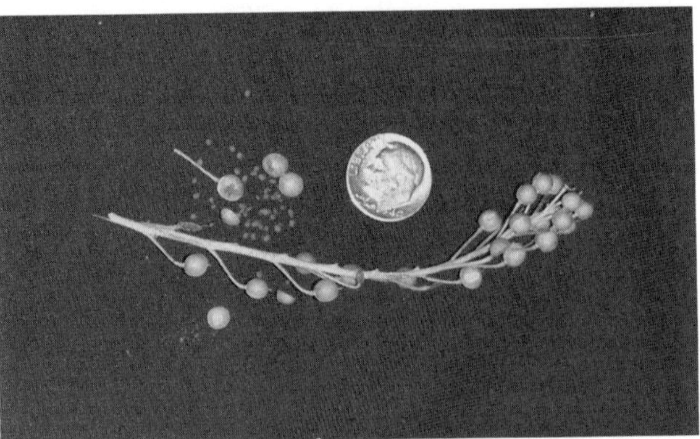

Top: Fig. 5.35. Bladderpod (*Lesquerella* spp.) in flower. This plant is an excellent source of commercially important hydroxy fatty acids.
Bottom: Fig. 5.36. Bladderpod (*Lesquerella* spp.) seed pods and the tiny seeds they contain compared to a U.S. dime. The seeds contain the valuable hydroxy fatty acids.

similar to the 18-carbon ricinoleic acid from the castor bean. This species will grow over a wide area from southwest Kansas, southwest Colorado, New Mexico, southeast Utah, west Texas, and western Arizona to north central Mexico. In southwest Texas it grows in massive populations in dry, open caliche soil. It is a fall-spring annual and exhibits good qualities for cultivation, including tolerance to heat, drought, and cold. Spring harvesting, usually in May, can be done by combining. One difficulty in harvesting is that rain causes immediate opening of the dried capsules and loss of seed, a problem known as hydrostatic dehiscence. *Lesquerella* species are polymorphic and exhibit a wide range of variation that offer the plant breeder adequate genetic material to overcome such problems as hydrostatic dehiscence.

Bladderpods can be grown with less water and on poorer soils than castor beans, and are free of castor bean's toxic and allergenic properties. Preliminary genetic and agronomic studies have been undertaken and seed yields of more than 1,800 pounds per acre have been obtained, which will produce about 540 pounds of oil. Recent experimental work on utilization has shown that when *Lesquerella* oil is polymerized in the presence of polystyrene, a new class of tough plastic is formed. The bladderpods provide the potential of a cash crop grown in desert regions that would not compete with conventional crops and could eventually offer a low-priced commodity for the chemical industry. For the U.S., if *Lesquerella* oil would replace imported castor oil, it would help the U.S. economy and its balance of trade problem. With these considerations in mind, a thirty-acre experimental plot of *Lesquerella* was established by the Univerity of Arizona in the fall of 1990. Photos of bladderpods in blossom and the seed pods are shown in figures 5.35 and 5.36.

In chapter 4 we discussed two food products produced by *Distichlis* species, WildWheat Grain and NyPa Forage. Non-food products are also produced.

NyPa Turf

NyPa Turf, the third *Distichlis* species ready for exploitation by NyPa, Inc., is a short turf variety (1–4 inches) for use as a lawn grass. This grass will grow in most any type of soil, but prefers heavy soils. Water salinity can vary from 1,000 to 10,000 ppm or greater. This species is unaffected by the heavy metals present in

Fig. 5.37. NyPa Turf, a variety of *Distichlis palmeri* that provides a luxuriant, weed-free, and easily-maintained lawn. Photo courtesy of NPY-NyPa Inc.

some water supplies. Turf plugs or sod are used in planting and 2–6 months are required for 100 percent ground cover, depending on the density of the initial planting. The salt present in the water eradicates most common weeds and the tight thatch requires little or no mowing. A typical lawn may go a month between mowings and irrigation. NyPa Turf (figure 5.37) does not produce pollen and thus should not be an allergen as are so many grasses. This selection should find wide application for home lawns, golf courses, parks, and highway media, especially where water quality, herbicides, or mowing costs are major budget items and ecological considerations are important.

NyPa Reclamation

NyPa Reclamation, the fourth NyPa *Distichlis* specialty, can be used as a reclamation grass. It is a perennial variety useful in stabilizing salty, heavy clay soils, mine tailings, and sewage outfalls where conventional plants cannot grow. Total ground coverage can be achieved in a year or two using approximately one pound of seed per acre. Because it is a hardy perennial, reseeding should not be required.

Agroforestry

Most of the arid and semiarid regions of the world are not only short of water, but fuel as well. Although so-called advanced areas of the world use petroleum or natural gas as the prime source for energy and fuel, most Third World countries depend on wood and/or manure almost exclusively. The combined effects of overpopulation, overgrazing, and desertification have created a great shortage of fuelwood and simultaneous destruction of the land base. A return to a system once employed widely in Africa called agroforestry holds promise of reversing the present trend. Experts now agree on the value of this age-old, integrated land use system.

Chapter 3 addresses techniques or systems which help to prevent soil erosion from wind and torrential rains as well as the need to increase the fertility and texture of the soil. Intercropping, the practice in which two or more crops are intermingled in the same field, and alley cropping, the practice in which food crops are grown between hedgerows of fast-growing leguminous trees, are both used for this purpose. The latter is a prime example of agroforestry. It provides protection from soil erosion and enhanced fertility through nitrogen fixation, and when the trees are pruned the leaves can be used as mulch or fodder. The stems provide fuelwood and stakes. Depending on the selection of the tree and its capacity for coppicing, poles too can be obtained.

The choice of trees for agroforestry, and alley cropping in particular, is very important. Those selected must have deep roots to avoid competition with food crops for water and nutrients. The species should be fast-growing, climatically tolerant, and vigorous nitrogen-fixers in order to provide the maximum utility for the soil, its herbaceous/animal crops and direct human needs such as fuel, stakes, and poles. The U.S. National Academy of Sciences identified over 1,200 species as firewood species, of which about 700 were given top ranking. This signifies that they are potentially valuable species that deserve increased research attention. Of these, 87 were described in some detail in two National Academy of Sciences volumes (1980, 1983) of a treatise on the topic.

Each species, of course, has its merits and shortcomings, its admirers and detractors. Two genera of note are the Linnaean and Prosopis. One species of each genus will be discussed; leucaena and mesquite, respectively.

Leucaena

Leucaena leucocephala (figure 5.38), one of the ten major species of *leucaena* (pronounced loo-see-nah) is perhaps best suited to the needs of arid and semiarid Third World countries with adequate water resources. It is a tropical leguminous tree that has a great variety of uses including fuelwood, charcoal, poles, a browse legume for animals, and edible young seeds and pods for humans. Its superb nitrogen-fixing properties make it an ideal organic fertilizer for intercropping and alley cropping and it simultaneously serves as a windbreak and erosion control agent. Its pest resistance and durability under grazing, cutting, fire, and drought have become legendary.

Leucaena species occur as native populations from Texas to Peru and from sea level to over 8,000 feet. It is now naturalized throughout the world's tropics between 30 degrees north and 30 degrees south latitudes on soils that are not too acidic and at elevations up to 5,000 feet. Under conditions where the winters are not too severe, leucaena can be grown to 35 degrees north and 35 degrees south latitudes. It prefers soils that are neutral to alkaline, but some species in the genus can tolerate acid soils. It does not tolerate soil conditions where flooding or waterlogging persist. It does best in fertile warm soils that are deep, friable, and moist. Leucaena is notably drought-hardy, tolerating annual rainfalls as low as 14 inches, but it prefers more than twice that amount. Its drought resistance is due in part to its strong, deeply penetrating root system and tiny leaflets. If it can reach adequate subterranean water it will survive nicely during the intense 6- to 9-month dry season of many tropical areas. Under these conditions leucaena is often the only legume-producing green foliage. The optimum temperatures for leucaena growth appear to be between 75 degrees F. and 86 degrees F., with little growth below 50 degrees F. It will survive frosts provided they are not too heavy and frequent.

Leucaena is a rapidly growing tree. Under good management transplants can reach three feet in one month, ten feet in three months and sixteen feet in five months. When cut to the ground

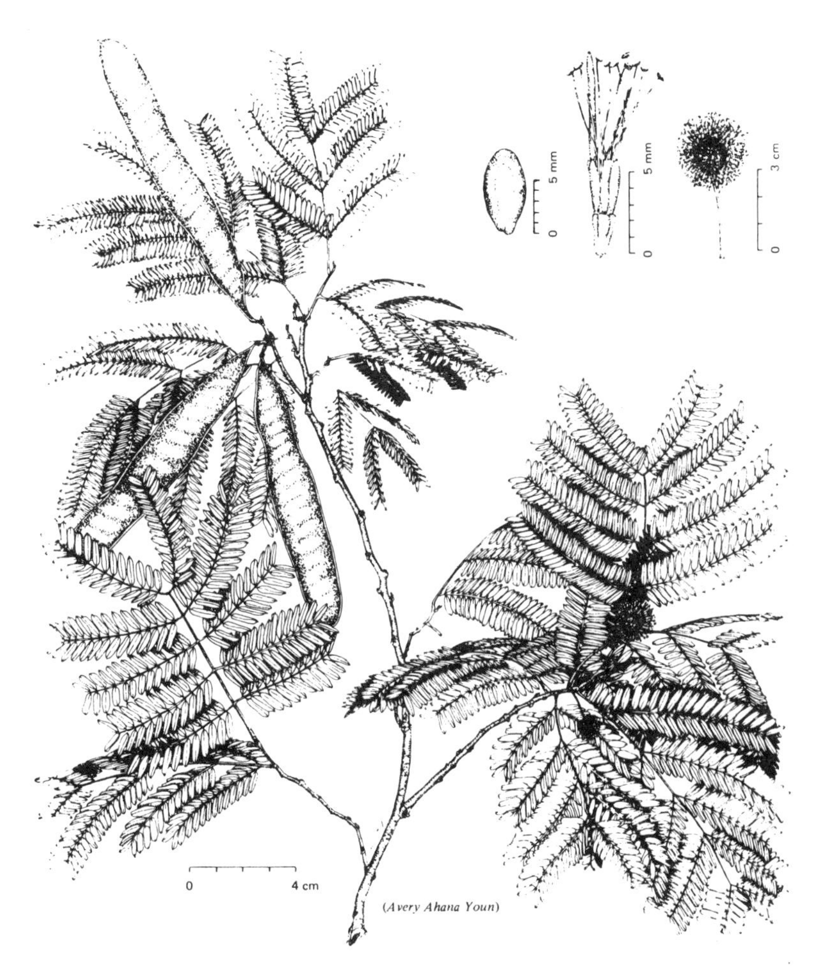

Fig. 5.38. Leucaena (*Leucaena leucocephala*). Note the small leaves characteristic of arid-adapted tree species. Photo provided by Dr. Donald L. Plucknett.

the tree can produce a cluster of new branches up to thirty feet in length in two years. The forage of leucaena provides protein yields among the highest reported for plants. Yields of forage, however, are highly light- and temperature-dependent. Harvest frequency must be varied during the year to maximize forage quality, with intervals ranging from 60 to 120 days, averaging 90 days. The tree can be grazed or cut back regularly to provide continuous high yields of a nutritious foliage that is favored by cattle, horses, pigs, sheep, rabbits, and other farm animals. The forage from this tree is so abundant and nutritious that its dried leaf meal now commands an international market as a protein and vitamin supplement for poultry and large animal feed. The protein value and amino acid fractions compare favorably with those of alfalfa, as does its overall yield on a dry ton per acre per year basis.

As with most legumes, leucaena forage produces some toxic effects in animals when ingested at very high levels. Most conspicious of these is loss of hair in nonruminants, but fortunately all toxic effects, for which the chemical mimosine accounts, are reversible.

In the future leucaena may well become a prime source of fuelwood for developing countries because its yield of hardwood can be comparable or superior to any known tree, a fact worth noting as many feel the fuelwood crisis in most Third World countries is as severe as the food crisis. Tree growth can reach 50 feet in five years with annual wood yields greater than 50 dry tons/acre/year, which is outstanding. The wood is hard, yielding a high-energy fuel and charcoal. It can also make a satisfactory pulp for kraft paper or newsprint. Also important is the fact that the trees can be harvested as posts, poles, and lumber. Leucaena is an excellent tree for reforestation of denuded, dry, nitrogen-poor soils abandoned after shifting cultivation, overgrazing, or wanton deforestation. If grown properly, leuceana could provide the ton or more of fuel required annually per person in those areas with minimal use of land and maximal ecological effectiveness. Under favorable conditions an acre of leuceana could produce enough fuel for at least four families. The wood has minimal bark, and burns well and cleanly with little smoke and low ash. It makes an excellent charcoal with a heating value over 70 percent of that of fuel oil. Charcoal is ideal for developing countries because it is simply made, easy to transport and store, and clean burning.

Perhaps the most important role of all for leucaena will be its use as a crop for the small farmer in developing countries. Today many farm families expend as much energy harvesting or searching for wood for fuel as in producing food. Leucaena could provide wood on site, plus poles, green manure, forage for the animals, and seeds and pods for the farmer's own table. The raw young seeds are a favorite snack for many people and contain over

30 percent protein. In addition, the wood compares favorably with others for artistic carving and thus for export. It is close-grained and is readily workable, lacking excessive gums, knots, and variations in heart and sapwood. The wood is yellowish-white, with heartwood somewhat darker.

The use of leucaena as a hedgerow for intercropping or alley cropping could be particularly important. When used for intercropping for animal grazing the hedgerows of leucaena generally will be spaced from 6 to 12 feet apart with an appropriate grass or grasses planted between rows. The trees provide important protein supplements to animals and some nitrogen to the associated grasses as well as shade and protection from erosion. The trees have also been shown to be effective in stopping or retarding grass fires. When used for alley cropping the space between rows is highly dependent on the choice of the other crop(s) to be grown and whether it needs support, e.g., pole beans. Also, regular pruning of leucaena side branches is desirable. This practice returns nitrogen to the soil equivalent to about 900 pounds of ammonium sulfate fertilizer per acre per year, assuming about 400 trees per acre. The added contribution by dropped leaves of phosphate, potash, calcium, and other salts is significant and leads to soil enrichment and improved texture.

Despite the tremendous advantages of employing leucaena as outlined here, caution must be exercised. A suitable geographic range for this plant is limiting, i.e., it does not tolerate freezing temperatures or waterlogging well and generally requires neutral or alkaline soils. Perhaps most limiting is its water requirements. Even though it is drought-resistant, it probably acquires this resistance by tapping subterranean water supplies. It is not a true xerophytic plant. Care must be exercised in selecting the right species for each anticipated site for use. Leucaena deserves serious consideration for many semiarid areas accompanied with well-conceived and executed experimentation before commitment to extensive plantings. Most locations selected will require supplemental water at least until the trees have become well established.

Mesquite

In areas that are truly arid and do not have a supply of irrigation water, one of the better trees to be considered is *mesquite* (figure 5.39). Even it, however, requires added water for good stand es-

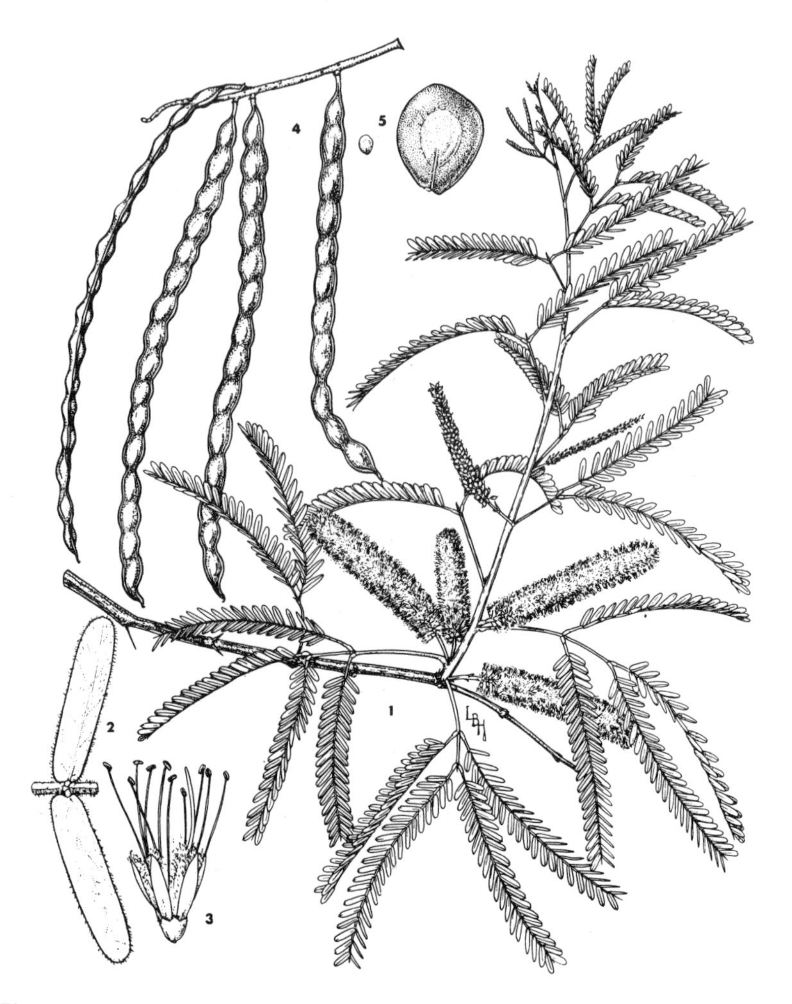

Fig. 5.39. Velvet mesquite (*Prosopis juliflora*). 1) branch with spines, bipinnate leaves, and spikes of flowers; 2) a pair of secondary leaflets, enlarged; 3) flower, showing the ten stamens and the separate petals, each hairy on the inner side of the apex; 4) fruits on the axis of an old flowering spike; 5) seeds, in two degrees of enlargement. Reprinted with permission of the University of Arizona Press, copyright 1991, Lucretia B. Hamilton, artist.

tablishment. There are forty-three species of mesquite indigenous to North and South America, Africa, and Asia, ranging in height from six feet to nearly fifty feet, and they exist in a variety of growth forms. Mesquite is a leguminous tree with unusually deep roots. It can provide firewood, food for either human or beast, nitrogen to deficient soils, and protection from erosion.

Mesquite naturally grows in some of the hottest and driest areas of the world; for example, Death Valley, California, where the daily average maximum temperature for July is about 113 degrees F., and where the average annual rainfall is less than two inches. Its secret lies in its ability to produce a deep-penetrating tap root. If groundwater is available within 35 feet of the surface the tree will survive and often flourish. Root depths of over 250 feet have been recorded. Trials designed to test the usefulness of mesquite for reforestation in Pakistan, where summer temperatures exceed 110 degrees F. and annual rainfall is less than ten inches, have been successful. One must conclude that mesquite is truly drought-resistant.

Various species of mesquite differ greatly in tolerance to freezing temperatures, ranging all the way from very frost-sensitive to extremely frost-hardy. Its growth range extends to at least 40 degrees north and 40 degrees south latitudes. Mesquite is also tolerant of poor soils, in part because it produces its own nitrogen, and is tolerant of poor-quality water. Studies have shown it will tolerate combined chloride and sulfate concentrations of up to 6,600 ppm. For perspective, water salinities of 1,000 to 3,000 ppm are suitable for irrigation of moderately saline-resistant crops. Few plants do well above 3,000 ppm, thus mesquite is very saline-tolerant and, accordingly, can be grown where few other crops are possible.

Although mesquite's ability to fix nitrogen has not been well quantified, a several-fold increase in soil organic matter and nitrogen content around and under mesquite trees has been observed in the United States and India. One study in Arizona showed that soils taken from under mesquite had a fourfold greater herbage yield than soils taken outside of mesquite's foliage cover. Increased soil organic matter under leguminous trees has been shown to provide increased soil cation exchange capability, increased soil water holding capability, and better soil structure.

Mesquite is most likely to be used in developing countries to provide a combination of benefits, i.e., erosion control, organic fertilizer, food for animals and/or humans, and fuel. Several stud-

Fig. 5.40. Small wild mesquite tree (*Prosopis* spp.) bearing mature fruit, Tucson, Arizona. Note that some pods have already fallen to the ground.

ies in India and Pakistan using a number of species have shown mesquite to be superior in controlling shifting sand dunes and wood production. It is safe to conclude that mesquite produces annual timber (clear bole) volume yields of over 1.5 dry tons/acre/year in areas of only 10 inches annual rainfall. Obviously, more rain produces larger yields. It is probably safe to assume also that the aerial parts, excluding the bole, and the roots, each contribute as much annual growth as the clear bole, thus the total dry matter yield is calculated to be at least 4.5 dry tons/acre/year. Generally speaking, the bole would be harvested for firewood and/or charcoal only as the trees pass their age of maximum pod productivity.

All mesquite species have indehiscent pods containing tightly bound seeds (figure 5.40). Depending on the species, the pods may be straight or curled, from 1 to 10 inches in length, and exist in a variety of colors. The pods contain about 13 percent protein and up to 30 percent sucrose. The seeds are generally small but contain about 27 percent protein. Pod yields of over 3 tons/acre/year have been reported growing in a Chilean salt desert existing solely on saline groundwater. Although this yield seems extraordinarily high, work in South Africa, Hawaii, and India tends to corroborate these data. South American mesquite varieties seem to be superior producers of both biomass and pods.

When Europeans first came to the New World they found mesquite pods to be a significant part, and in some cases a dominant staple, of many Indian diets. Among the Indians of the American southwest many tribes, and particularly those along the Colorado and Gila rivers, used mesquite pods as their chief source of food. The immature juicy pods were crushed and the resulting extract was drunk throughout the summer. The immature pods of sweet mesquite trees are similar in taste to immature pea pods or snow peas. Fermentation of ground pods of at least one species of mesquite produces a beer that was highly prized by some American Indian tribes. The dried mature pods can be broken into small pieces and eaten directly, or ground into a meal or flour, while the seeds are most often discarded. The flour is the starting material for a variety of dishes, but often it was dampened and formed into balls or cakes and stored for future use. Mesquite pods still serve as a food source among some Indian tribes, but are more often used as animal feed.

In terms of yield of feed per acre mesquite is outstanding in hot, dry climates. Mature crushed mesquite pods have been reported to contain about the same nutritional quality as maize as a livestock feed, and the pods are relatively easy to collect and store. To obtain the maximum benefit as a livestock feed, the pods must be ground. This assures the release of seed protein which otherwise pass undigested through the animal's alimentary tract. Mesquite meal should not be used as the entire ration, however, because it causes constipation. Some mesquite species also contain trypsin inhibitors and hemaglutinins, which when present in too high a concentration tend to accumulate with deleterious effects. This is not unusual as these compounds are present in most legume seeds. These ill effects are largely overcome by supplementing the diet with hay or other fodder.

In conclusion, mesquite has already proven its value in several arid regions and deserves serious consideration for more widespread use. Care must be exercised to ensure that the proper species has been selected for any particular site. Although mesquite species are generally free of pathogens, there are exceptions. Then, too, some species are very aggressive and become difficult to confine, and pod and/or wood yields vary considerably from species to species. The genus *Prosopis,* to which mesquite belongs, is extremely variable in growth habit and rate, pod and leaf morphology, and site adaptability to salinity, heat, and cold. Different species must be selected depending on the climatic and soil conditions of the site as well as the most desired characteristics, i.e., high pod yield, high protein pods, woody mass for firewood use, etc. Failure of one mesquite selection to fulfill a particular use or ecological niche should not lead to abandonment of the species. Mesquite must be regarded as a multiple-use crop with both short-term and long-term benefits for arid lands agriculture and the ecology.

Other Possibilities

It is beyond the scope of this book to give a complete listing of all potential new crops that could provide products or items of commerce other than food. There are many and the products are varied so only three have been selected for discussion to illustrate the point. Each is native to an arid region and has been successfully grown in Arizona and California.

Gum tragacanth, Astragalus (several species of the group tragacantha), are collected from wild bushes (see figure 5.41) in the arid mountains and plateaus of several countries in or near the Middle East, with Iran as the largest producer. The plants are widely scattered in remote, inhospitable semideserts, but the natural stands have all been depleted. Gum tragacanth is highly prized and finds uses in pharmaceuticals, cosmetics, and as a thickening agent in foods. The gum (figure 5.42) commands high prices ($20—35 per pound in 1989). No other gum has been found that is a complete substitute for gum tragacanth. Dr. H. Scott Gentry received support from the National Science Foundation to investigate this plant, and his work has been encouraging. Gentry believes we can mechanize tapping and gum collection, thereby largely eliminating expensive hand labor. It appears that gum tragacanth could be cultivated to yield another crop for selected arid or semiarid sites.

Top: Fig. 5.41. Azerbaijan, Iran. *Astragalus* spp., the source of gum tragacanth (photo by H. S. Gentry).
Bottom: Fig. 5.42. The spiraling exudate of gum tragacanth, Isfahan, Iran (photo by H.S. Gentry).

Fig. 5.43. Red squill (*Urginea maritima*) plantation located near Riverside, California. Just entering the dry season, the leaves are becoming desicated and will not regenerate until the following rainy season. If mature enough, however, the bulb will send forth a flower spike during the dry season. Photo provided by Dr. A. J. Verbiscar.

Fig. 5.44. Red squill (*Urginea maritima*) can be grown for its scilliroside, used as a rodenticide or for its flower, as illustrated in this photo.

Red squill, Urginea maritima, is native to the Mediterranean region and was used in medicine by the early Greeks. Later it was used as a rodenticide throughout Europe and exported to the United States for the same purpose. During World War II an attempt to establish the plant in Mexico and the United States was made to assure a continuing supply of the rodenticide. Most work on red squill was halted at the end of the war. However, Dr. Gentry and his colleagues continued their efforts to select the best strains and propagated clones for sustained high yield of scilliroside. Scilliroside, a glycoside, is the primary toxic component in red squill. Good progress was made but, alas, red squill has fallen from favor as a rodenticide, at least for the moment, so no market for scilliroside now exists. But the story does not end here.

Red squill is a perennial bulbous plant that grows in arid climates. It grows through winter and spring when it is cool and moist and stops growing completely with the first hot days of summer. The leaves thoroughly dry out and the bulb goes into dormancy, unaffected by the hot, rainless months of summer. The first flowers begin to appear in the sixth year as tall attractive spikes during August and September. These are in much demand by florists because they are so unusual and attractive, and they last about two weeks. They wholesale for $1.00 to $1.50 each depending on size and shape. Red squill grows well in California (figures 5.43 and 5.44) near coastal areas receiving 12 to 20 inches of rainfall annually or on well-drained soils in Arizona with minimal irrigation. Red squill requires full sun for best growth and withstands temperature extremes from 20 degrees F. to 115 degrees F. without serious injury. Once established, the bulb appears to grow for at least twenty-five years and probably much longer.

It seems likely that a new floral crop is at hand. Although five to seven years must elapse between investment and first revenue, maintenance costs are minimal. From 10,000 to 12,000 bulbs per acre appears to be appropriate with each averaging one flower annually starting with year six. In fact, one grower is already established.

One of the most interesting potential new products really is not new at all. It was in great demand for over 300 years and then within just a few years the demand nearly ceased. It is still produced even today, however in limited quantities. The product is *carminic acid,* a strong and color-fast material dye. It was superseded by synthetic aniline dyes in the latter part of the nineteenth century because they were so much cheaper.

When the Spanish conquered the Aztecs and Incas in the early part of the sixteenth century, they discovered that these Indians had a brilliant red dye which was much superior to any known in Europe. It was so much brighter and more intense than the red dyes of the day that it became an instant success, and the Spanish

Fig. 5.45. The small insect called cochineal (*Dactylopius* spp.) attaches itself to a prickly pear cactus, hides under the cottony white material, and produces carminic acid, a brilliant red dye. Using a pen knife, the cottony material has been partially removed to expose the insect.

became Europe's only source of *cochineal dye* (carminic acid) for nearly 200 years. Carminic acid is produced by the scale insect, cochineal (*Dactylopius spp.*) that lives on the prickly pear cactus (*Optunia spp.*).

As warm weather comes to the desert, patches of white cottony material appear on some pads of the prickly pear. Beneath this fluffy mass hides the cochineal female (figure 5.45), permanently attached to the plant by sucking mouth parts inserted into the cactus. She contains carminic acid—up to 10 percent of her body weight. The male is much smaller, has wings but no mouth parts and after mating soon dies of starvation. Harvesting is accomplished by scraping the creatures from the pads and immersing them in hot water or drying them in the sun. The males, which contain no dye, are removed by sieving. It takes about 70,000 insects to produce a pound of carminic acid. The dye is still very expensive because its production requires much hand labor. However, it may still make a strong comeback because of its uses in medicinal tracers, color photography, microscopy stains, and particularly as a colorant in cosmetics, pharmaceuticals, and foodstuffs. In the United States, it is now the only natural red food coloring authorized by the Food and Drug Administration. The industries that use this dye could make cochineal farming important once again.

It would be inappropriate to close this chapter without at least mentioning the great potential for producing salt-tolerant ornamental plant species. Germ plasm of over 1,000 halophytes have been collected by the Environmental Research Laboratory of the University of Arizona alone. Many of these plants are appropriately sized for ornamental use and are strikingly beautiful. Use of them for beautification of home and community environments holds the potential for enhanced surroundings while dramatically reducing the use of precious fresh water.

Many arid regions possess saltwater aquifers or are near seas. Fresh water is scarce and expensive. The use of NyPa Turf and halophyte ornamentals could dramatically improve the quality of life and surroundings inexpensively. These plant species present a challenge and opportunity for both entrepreneurs and arid-land located communities.

SELECTED INFORMATION SOURCES

Adams, R. P. 1986. "Bio-Renewable Sources of Chemicals and Fuel in the Chihuahuan Desert." *Chihuahuan Desert—U.S. and Mexico* 11:131–149.

Barclay, H. S., H. S. Gentry, and Q. Jones. 1962. "The Search for New Industrial Crops II: *Lesquerella* (Cruciferae) as a Source of New Oilseeds." *Economic Botany* 16(2):95–100.

Beckett, R. E., R. S. Still, and E. N. Duncan. 1938. *Rubber Content and Habits of a Second Desert Milkweed* (Asclepias erosa) *of Southern California and Arizona.* Technical Bulletin no. 604. Washington, D.C.: United States Department of Agriculture.

DeVeaux, J. S., and E. B. Shultz, Jr. 1985. "Development of Buffalo Gourd (*Cucurbita foetidissima* HBK) as a Semi-Aridland Starch and Oil Crop." *Economic Botany* 39:454–472.

Doten, S. B. 1942. *Rubber from Rabbitbrush* (Chrysothamnus nauseosus). Bulletin no. 157. Reno: University of Nevada Agricultural Experiment Station.

Goldstein, B. 1988. "Technical and Economic Feasibility of Buffalo Gourd as a Novel Energy Crop." New Mexico Research and Development Institute report NMRDI 2-72-4213.
Hinman, C. W. 1984. "New Crops for Arid Lands." *Science* 225:1445–1448.
Hinman, C. W. 1986. "Potential New Crops." *Scientific American* 255:32–37.
Hinman, C. W., A. Cook, and R. I. Smith. 1985. "Five Potential New Crops for Arid Lands." *Environmental Conservation* 12:309–315.
Hogan, L., and W. P. Bemis. 1983. "Buffalo Gourd and Jojoba: Potential New Crops for Arid Lands." *Advances in Agronomy* 36:317–349.
Johnson, J. D., and C. W. Hinman. 1980. "Oils and Rubber from Arid Land Plants." *Science* 208:460–464.
Jolliff, G. D. 1984. "Jojoba Oil Faces Competition from Meadowfoam." *Biomass Digest* 6(7):7.
Knudsen, H. D. 1989. Personal communication.
Lucas, K. 1988. "Growers in 1988 to Produce 2–3 Million Pounds of Seed." *Jojoba Happenings* 16(November-December):1–3.
Morgan, R. P., and E. B. Shultz, Jr. 1981. "Fuels and Chemicals from Novel Seed Oils." *Chemical and Engineering News* 59(7):69–77.
National Academy of Sciences. 1977. *LEUCAENA: Promising Forage and Tree Crop for the Tropics.* Washington, D.C.: National Academy of Sciences.
National Academy of Sciences. 1984. *Firewood Crops: Shrub and Tree Species for Energy Production.* Washington, D.C.: BOSTID Publications.
National Research Council. 1985. *Jojoba: New Crop for Arid Lands. New Material for Industry.* Washington, D.C.: National Academy Press.
Ostler, W. K., C. M. McKell, and S. White. 1986."*Chrysothamnus nauseosus*: A Potential Source of Natural Rubber." In E. D. McArthur and B. L. Welch, compilers, *Proceedings, Symposium on Biology of Artemisia and Chrysothamnus,* pp. 389–394. USDA Forest Service General Technical Report INT - 200, Ogden, Utah.
Perdue, R. E., Jr. 1981."*Vernonia galamensis* (Cass.) Less. An Unappreciated Source of Epoxy Fatty Acid." Paper presented at joint meeting of American Society of Pharmacognosy and Society of Economic Botany, Boston, Mass., July 13–17, 1981. Abstract no. 54.
Perdue, R. E., Jr. 1988. Personal communication.
Perdue, R. E., Jr., K. D. Carlson, and M. G. Gilbert. 1986."*Vernonia galamensis,* Potential New Crop Source of Epoxy Acid." *Economic Botany* 40(1):54–68.
Princen, L. H. 1979. "New Crop Developments for Industrial Oils." *Journal of the American Oil Chemists Society* 56(19):845–848.
Princen, L. H. 1983. "New Oilseed Crops on the Horizon." *Economic Botany* 37(4):478–492.
Shultz, E. B., Jr., W. G. Bragg, D. L. Duke, and B. H. Malkani. 1989. "The Rootfuel Alternative to Scarce Woodfuel in Third World Drylands: Combustion Tests and Socio-Economic Studies." Paper presented at the

Conference on Energy from Biomass and Wastes XIII, sponsored by the Institute of Gas Technology, February 13–17, 1989, New Orleans, La. Published in the Conference Proceedings.

Skogman, H. 1979. "Effluent Treatment of Molasses-Based Fermentation Wastes." *Processes in Biochemistry* 14(1)5–6, 25.

Sperling, L. H., and S. K. Dirlikov. 1990. "Vernonia Oil Shows Promise as Reactive Monomer." *Chemical and Engineering News* 62:May 7.

Sperling, L. H., and J. A. Manson. 1983. "Interpenetrating Polymer Networks from Triglyceride Oils Containing Special Functional Groups: A Brief Review." *Journal of the American Oil Chemists Society* 60(11):1887–1892.

Timmermann, B. N., and J. J. Hoffmann. 1985. "Resins from *Grindelia*: A Model for Renewable Resources in Arid Environments." In G. E. Wickens, J. R. Goodin, and D. N. Fields, eds., *Plants for Arid Lands. Proceedings of the Kew International Conference on Economic Plants for Arid Lands, Royal Botanical Gardens, Kew, England, 23–27 July 1984*, pp. 357–367. London: George Allen & Unwin.

Watson, W. C. 1989. "Future Holds Many Opportunities." *Jojoba Happenings* 17(January-February):3–6.

Weber, D. S., T. D. Davis, E. D. McArthus, and N. Sankhla. 1985. "*Chrysothamnus nauseosus* (Rubber Rabbitbrush): Multiple-Use Shrub of the Desert." *Desert Plants* 7(4):172–180, 208–210.

Whitworth, J. W., and E. E. Whitehead, eds. 1990. *Guayule Natural Rubber.* Tucson: Office of Arid Lands Studies, University of Arizona.

SIX

Products, Coproducts, and Processes

Axiom: Unlike fossil fuels, use of biomass as an energy source cannot contribute more carbon dioxide to the atmosphere than it extracts from it.

In order to make more clear how new products and coproducts are derived from arid-land plants and how the bagasse (residual material following removal of primary product[s]) can be utilized, several processes will be described. Each plant discussed in chapter 5 will be considered where a process beyond that already given is required to secure the primary product and coproducts. Various uses and/or treatment of bagasse will be considered separately.

Seed Oil Extraction

Vernonia, bladderpod, buffalo gourd, and jojoba seeds are extracted for their primary product (oil) in much the same manner as other seeds, such as soybean, cotton, sunflower, and safflower, are extracted for their oils. Seed oil extraction has been commercially practiced for decades and usually involves a primary pressing or expelling operation to remove most of the oil followed by solvent extraction to remove the residual oil. Initial pressing is usually required if the seed contains a high percentage of oil because solvent extraction is most efficient for oil percentages of 15 percent or lower. To increase the efficiency of solvent extraction the solid material from the press or expeller is usually made into flakes not unlike breakfast cereal flakes. A number of different solvents can be used, but hexane is perhaps the most popular. The solvent must be recovered from both the extracted oil and the residual meal, usually by means of a desolventizer/toaster and solvent stripping/distillation system. All these techniques are well known, but modified to maximize the recovery of the particular seed oil under consideration, for each seed species has its peculiarities.

Vernonia galamensis seed contains about 40 percent oil of which approximately 80 percent is vernolic acid (*cis*-12,13-epoxy-*cis*-9-octadecenoic acid). This natural epoxy acid has industrial potential in PVC plastics, improved epoxy resins, other polymer blends, baked coatings, and reactive diluents for paints and coatings systems. Using a system as described above, one is able to recover about 30 percent of the seed weight as vernolic acid which has a ready market. One peculiarity of vernonia seed is that it requires heating at 195–200 degrees F. to inactivate a lipase before pressing and solvent extracting. This is not uncommon, however. Both soybeans and rapeseed, for example, must be treated to inactivate enzymes. Once the vernolic acid is removed, one must consider the residual material which constitutes about 60 percent of the initial seed weight.

Fortunately, the defatted flakes are rich in protein (44 percent). Methionine was found to be the first limiting amino acid, with lysine the second. Carbohydrates comprise about 7 percent of the total. The levels of minerals such as calcium, potassium, phosphorus, and magnesium not only meet nutritional requirements, but are higher than those of contemporary oil seeds. Feeding trials suggest that vernonia-defatted flakes have potential as a new rich protein and mineral source for animal feed, thus making salable all parts of the harvested *Vernonia galamensis* seed.

Bladderpod (*Lesquerella* spp.) seed contains about 30 percent oil, of which at least 60 percent is lesquerolic acid. This acid is similar to ricinoleic acid from the castor bean and can be a substitute for it in most applications, including greases needed for both industrial and military applications. In addition, when bladderpod oil is polymerized in the presence of polystyrene, a new class of tough plastic is formed. The pressing and solvent extraction procedure to obtain bladderpod oil is essentially the standard seed oil extraction procedure. This oil, and lesquerolic acid when required, are the primary products. The remaining defatted flakes, which constitute some 70 percent of the total seed weight, resemble rape and crambe meals and thus will probably find use as an animal feed.

Buffalo gourd (*Cucurbita foetidissima*) seed contains as much as 40 percent of its weight as a high-grade edible vegetable oil. It can be extracted using standard seed oil procedures. The defatted residual meal is nearly 50 percent protein with a good amino acid profile. Being similar to the meals of other seed oils, it will probably find a useful role as an animal feed supplement. Accordingly, the entire buffalo gourd seed is expected to be salable.

Jojoba (*Simmondsia chinensis*) oil production procedures differ somewhat from normal seed oil extraction in that most, but not all, is obtained without a solvent extraction step. The processor first must dehull the seed before pressing commences. Since the seed is approximately 50 percent oil by weight it must be pressed several times to retrieve 90–95 percent of the total oil. Each time the meal is pressed it is done so at higher pressure. The more pressure applied the poorer the quality of oil produced. Although this pressing procedure is referred to as "cold pressing," each subsequent pressing at higher pressure results in an elevated temperature, which produces contamination of the expressed oil. Jojoba oil is actually a liquid wax. As obtained from the expeller, it is often cloudy and discolored. This is overcome by bleaching with activated clay and filtering, which is common treatment with seed oils. In practice, the oil is usually passed through a plate and frame filter or similar filtering device several times, then pasteurized just before packaging to ensure sterility. Several jojoba processors obtain their oil by a combination of pressing and solvent extraction. When the combination is used the seeds are generally cold pressed only once, leaving considerable oil in the meal. This meal is then solvent-extracted to remove the rest of the oil. Using the combination of pressing and extraction yields more total oil from the seed and usually results in better quality oil. It should be noted that solvent extraction is made available on a custom basis to processors who lack solvent extraction equipment in order to obtain as much oil as possible.

The industry has established a number of standards to assure uniformity and appropriate grading of the products produced. Jojoba oil is sold not only as a pure oil, but as chemically altered derivatives as well, such as a transesterified jojoba oil and wax, a hydrogenated jojoba wax, and a microencapsulated wax. Most jojoba oil and derivatives are presently sold to the cosmetics industry, although some sulfurized jojoba oil is formulated to produce automotive lubricants.

Unfortunately, the residual jojoba meal has but limited utility at the moment. Some is being sold as a radiator "stopleak" to the automotive industry and a small quantity to the cosmetics industry as an abrasive for natural scrub. The defatted seed meal contains about 25 percent crude protein but cannot be used as a feed supplement because it also contains anti-nutritional factors, simmondsin and its derivatives. In addition, the meal contains significant amounts of tannins which most animals find offensive to the taste. Research at several locations has addressed the simmondsin problem and several different processes have been developed, one of which, from the USDA Northern Regional Research Center, seems particularly promising. Water is added during the solvent (hexane) extraction process, allowing enzymes inherent in the jojoba seed to biodegrade the simmondsins and thereafter, following the removal of the hexane, the meal can be dried to yield a safe animal feed. Another Northern Regional Research Center process uses fermentation to destroy the simmondsins. It may be even more promising and has been scaled up to the pilot plant level. The tannin problem can be overcome by using jojoba meal as a supplement to other feeds, thus reducing the overall tannins level below that which animals find objectionable. By utilizing all the defatted jojoba meal produced, the economics of jojoba processing would improve significantly.

It should be noted that in the past some seed oils, e.g., soybean oil, have been evaluated as possible liquid fuels or fuel extenders for internal combustion engines. Most emphasis has been given to using seed oils as diesel fuels. To date, it appears that seed oils are not well suited to this use. In all probability the oils derived from vernonia, bladderpod, buffalo gourd, and jojoba will not find a practical application as fuels either.

Whole-Plant Extraction

Unlike removing oil from seeds there is little commercial precedent for extracting products from whole plants. Perhaps guayule serves as the best example. There are significantly more problems in dealing with whole plants than seeds. Whole plants are not uniform physically like seeds, leading to handling problems, nor is the concentration of the desired primary product as high. In fact, in considering guayule, rabbitbrush, and grindelia, none normally contains a concentration of primary product of even 15 percent of the total weight of the material to be processed. Compare this to the seeds containing about 30–50 percent. Accordingly, considerable bagasse must be handled and appropriately dealt with to find coproducts and/or utilized economically or the financial and environmental consequences can be disastrous. Even harvesting whole plants is much more difficult than harvesting seeds.

The history of obtaining rubber from guayule (*Parthenium argentatum* Gray) will illustrate the point. Over the years there have evolved succeeding processes, each rendering higher yields and better quality rubber as well as development of coproducts. The preferred guayule extraction process is probably applicable to rabbitbrush with only minor changes and is similar to that used to extract resins from grindelia.

Unlike *Hevea,* which produces rubber latex that is easily obtained from the tree by tapping, guayule stores its rubber within the cells of its bark and woody tissue. The cells must be ruptured to obtain the rubber. The very first method of accomplishing this was with human teeth. Later, various kinds of mills were used. All current procedures first grind the plant material and then subject the resulting coarse particles to a combination of compression and sheer force employing equipment like that used in oilseed processing.

Over the years three basic commercial approaches to isolate guayule's rubber have evolved: flotation, sequential extraction, and simultaneous extraction. In the flotation procedure the shrub was first parboiled and defoliated, followed by milling to produce coarse chunks or chips. These were agitated with a dilute caustic solution which yielded rubber "worms." The "worms" also contained much resin so they were collected and deresinated by extraction with acetone. The resulting crude rubber was dissolved in hexane, antioxidants added, and the solution filtered to remove all insoluble material such as dirt. The rubber was then isolated by steam desolventization. This method produced rubber satisfactory for most uses, but contained a higher than desired amount of resins. The most undesirable feature of the method is its production of large volumes of caustic flotation media. This effluent contains substantial levels of dissolved salts, plant extracts, and sodium hydroxide, and thus cannot be safely discharged into the environment without extensive treatment.

Sequential extraction addresses both major problems of the flotation process: residual resin in the rubber and an undesirable process effluent. In this process the defoliated, ground, and flaked shrub is first treated with a solvent, such as acetone, to remove the resins present, and then with a second solvent, such as hexane, to remove the rubber. Antioxidants are generally added to the solvents to stabilize the rubber. The sequential extraction process provides good yields of high-quality resin-free rubber. The process is amenable to continuous operation, versus batch operation, and can use commercially available equipment. The most important shortcoming of the process is solvent recovery. In any extractive process, incomplete recovery of solvent from residual solids is a primary cause of solvent loss. Having two extraction steps, this process is more prone to solvent loss than a process with only a single extraction step.

The simultaneous extraction process has but one extraction step, and thus potentially can be run with less loss of solvent than the sequential method. The choice of the primary solvent or solvent mixture is extremely important, however, for it must efficiently extract both rubber and resin, yet readily yield pure rubber upon the addition of another solvent which precipitates it. Then it must be easily and completely recovered from the biomass. In addition, it should be easily separated from the rubber-precipitating solvent. As implied, the simultaneous extraction process treats finely ground or flaked guayule shrub with a solvent or solvent mixture that very efficiently extracts both rubber and resin. The resultant solution is treated with another solvent which is miscible with the first but coagulates the high-molecular-weight rubber. By judicious choice of sequential additions of solvent, amount of solvent, and temperature, a number of rubber fractions of varying molecular weights can be obtained, resulting in purer products that can meet discrete product specifications. Once desolventized, the resin too can be fractionated by mixing with an organic solvent in the presence of a highly polar medium such as water. The simultaneous extraction process provides for the efficient production of the primary product, rubber(s) plus several coproducts, as discussed above in chapter 5.

Rabbitbrush (*Chrysothamnus* spp.) is a particularly interesting plant for it produces rubber, useful resins, and holds potential as a biomass energy source. Accordingly, it can yield all three, and it is highly probable a modified simultaneous extraction process such as developed for guayule will be used to produce them.

To obtain the primary product, resin, from *Grindelia camporum* requires a much simpler process than that required by guayule. The resin is most concentrated in and around the grindelia flower and secondarily on the surface of the leaves and stems of the plant. Since the primary product is on the plant surface it need not even be ground. It can be removed by washing or soaking the whole plant in an appropriate solvent or solvent mixture in the simplest of equipment. The resin is desolventized and used as a crude resin or may be purified by use of multiple solvents, reminiscent of the guayule process. The resin so obtained can be made to meet various specifications.

The bagasse from this process can be chopped or ground and utilized as an animal feed or it may be further treated to produce other coproducts. Grindelia bagasse differs from most bagasse in this respect. Of course, it can also be treated as ordinary biomass, for once the extractables are removed it is quite similar to the bagasse from most species, whether they are plants grown for a specific product, e.g., seeds, rubber, resins, lumber, or as in the case of some trees, just for fuel. Accordingly, it is important to consider alternative uses for bagasse/biomass.

From the time humans first began to use fire they were virtually limited to wood as their source of fuel until the early 1800s when coal and peat gained limited use. In much of the world, and particularly in Third World arid regions, wood and/or manure are still the main sources of fuel. Even as recently as 1850, wood accounted for 85 percent of U.S. fuel, although coal was gaining acceptance rapidly. By 1875 coal use accounted for about one-third of the fuel supply, while petroleum and natural gas were just being introduced. In terms of total energy production, by the turn of the century coal use in the U.S. was up to 75 percent, wood down to about 15 percent, petroleum and natural gas about 7 percent, and all other forms of energy accounted for the small remainder. The latter group was, and is, primarily hydropower, although solar and wind are gaining. By 1950 petroleum and natural gas accounted for half of all the U.S. energy consumed and by 1975 the margin had grown to over 75 percent. United States dependence on oil and natural gas is cause for great concern, particularly when at least half of the oil it uses must be imported, even more now than before the oil embargo of the 1970s, and its reserves are declining at an ever-accelerating rate. The growing burden of imported oil accounted for about one-third of the United States' 1988 trade deficit.

The quantities of oil produced and used daily is so immense it is almost impossible to comprehend. For example, the U.S. alone uses over seventeen and a quarter million barrels (42 gal/bbl) a day. This is roughly equivalent to consuming the total contents of a three-foot-diameter pipeline which stretches from New York to San Francisco each and every day of the year. Thus it is understandable that the rate at which oil reserves are being depleted is alarming. For example, a 1977 U.S. government report stated that "half of all the oil that has ever been produced has been taken from the earth in the past 10 years." That was just the ten years between 1967 and 1977. As another indication of U.S. oil reserve depletion a noted authority, Earl T. Hayes, in January 1979 was quoted as saying, "In the 1950s we found one and a quarter barrels of oil for every barrel we extracted, but by the late 1970s this has dropped to one half a barrel." Drilling is even less productive today.

During the 1980s the U.S. and world reserves position continued to decline. Although the U.S. is, by far, the largest consumer of oil, by 1988 it controlled only 5 percent of the world's reserves. The OPEC nations, by contrast, use relatively little oil, but control over 70 percent of the world's reserves. Another problem is that reserves can be exploited, and become almost depleted before the event is noticed in the marketplace. At present consumption rates, the U.S. and world oil reserves are predicted to become depleted within a few decades, some saying as early as the year 2020 for the U.S. and 2040 for the world. Natural gas reserves are expected to become exhausted about the same time, although coal reserves are projected to last at least another 500 years. Obviously, there will be no single depletion date, for it is impossible to predict with precision either the rate of future use or discovery of new oilfields, but the reserves are finite and it is clear that they will be depleted relatively soon. Then, too, as oil production capacity begins to lag behind demand, the cost will begin to escalate and alternate sources of energy will be sought and more conservation methods will be utilized.

During the oil embargo of the 1970s the price of oil increased from about $5/bbl to over $40/bbl. Since the U.S. consumed more oil than it could produce and the price per barrel had escalated so rapidly, the federal government, some state governments, and several private companies all initiated research programs to find alternate sources of fuel to replace fossil crude oil. Some of the research focused on the conversion of biomass to liquid fuels through the production of alcohol or synthesis gas (syngas). Several projects sought to find plant species that produced hydrocarbons directly. Arid lands were looked upon as likely places to find hydrocarbon-containing plants, because these areas usually have a high incidence of solar radiation. To produce hydrocarbons, plants must first produce carbohydrates with subsequent chemical reduction to hydrocarbons. The plants' only source of energy to make this conversion is the sun. A few plants were indeed found which produce a "biocrude oil," and which can be catalytically cracked like crude oil to produce gasoline and other liquid fuels. Indeed, all of these were arid land-adapted plants. The projected cost of this "biocrude oil," without plant species improvement, was about $90/bbl. At the time that cost seemed reasonable, for there were those who predicted the price of crude oil could easily reach $100/bbl by 1990. Of course, what actually ensued was an oil glut and tumbling prices. Although the price of crude oil never retreated to its previous low, it did stabilize at $15–$20/bbl. With oil once again so abundant and inexpensive, interest in and support for research efforts to replace it almost vanished. Many projects were completely abandoned, but fortunately some were continued, albeit at a much reduced level. However, the results achieved in those few years provided the basis for many of the emerging new technologies which now make biomass an increasingly interesting and important raw material.

However, crude oil is still the most important, abundant, and inexpensive raw material ever known. Its virtues are many. Unfortunately, its supply is finite and will soon be exhausted. It is an extremely concentrated source of energy, easily and cheaply transported, and readily converted into alternate forms of mass. It is the very foundation of western material life. It is such an inexpensive and versatile substance that, through chemistry, it not only serves as a fuel, but as a raw material for many of the products once derived from natural plant sources, such as resins, dyes, adhesives, and discrete chemical compounds. Today, natural products survive only when their molecules are either too complex or too expensive to be produced synthetically from petroleum. This will—and is beginning already—to change. It is inevitable that as known crude oil reserves diminish and the cost of finding and extracting new oil deposits escalates, the price of crude oil will rise. At the same time much progress has been made in lowering the costs of converting various biomass sources into products and usable energy. Time and circumstances are rapidly making biomass-derived specific chemicals, rubber, resins, liquid fuels, and other forms of energy economically attractive and environmentally necessary.

The "greenhouse" effect is seriously disputed by some scientists, although most believe it to be real. If the earth is truly warming, the primary reason is most often attributed to ever-increasing levels of carbon dioxide in the atmosphere. Fossil fuels—coal, oil, and natural gas—are the primary culprits. By burning these we have released into the atmosphere in just a few decades the carbon dioxide which plants extracted from the atmosphere over millions and millions of years. This massive release, coupled with reckless reduction of our planet's ability to absorb carbon dioxide through destruction of its forests and ranges, are the primary causes of the greenhouse effect. We must promote repopulation of our earth with plants to absorb more carbon dioxide and simultaneously find sources of fuels and raw materials other than fossil fuels. Plants/biomass can serve as such an alternate for a plant's only source of carbon is the carbon dioxide in the air surrounding it. When the plant is burned it cannot possibly release to the atmosphere more carbon dioxide than it extracted from it. In fact, when plants such as trees are harvested up to a third of the plant's carbon remains in the ground in the roots. Thus, substituting biomass for fossil fuels can only diminish atmospheric carbon dioxide levels, and thus reduce the threat of global warming. Additionally, biomass is low in sulfur-containing compounds compared to fossil fuels, thus another reason for its use as a cleaner energy source.

Of course, biomass can be burned directly to produce heat which is used for cooking and warmth in Third World countries, or used to produce process steam and/or electricity in more developed countries. However, it is a much more versatile and useful material than just a source of BTUs. Once the primary product and coproducts, if any, have been removed from the plant, the remaining bagasses are quite similar regardless of the plant species origin and all are reminiscent of wood. As such, once the extractables have been removed, the remaining biomass is composed almost entirely of cellulose, hemicellulose, and lignin. Although the percentage composition varies somewhat from species to species, in general, cellulose accounts for about 50 percent of the weight with hemicellulose and lignin sharing almost equally the other 50 percent. This biomass material is often referred to as lignocellulose.

Cellulose is a high-molecular-weight linear polymer of glucose, nature's most abundant sugar. Glucose is a six-carbon sugar easily digested or fermented in its free state. Hemicellulose is also a polymer which upon hydrolysis yields both five- and six-carbon sugars. The major sugar is xylose, a five-carbon sugar. Lignin too is polymeric, but much more complex in that it is a three-dimensional network of high-molecular-weight phenolic compounds. Lignin can be visualized as the cement that holds the plant's physical structure of cellulose and hemicellulose together. These biomass components provide a large number of varying uses and products depending on how the mass is treated or how each component is treated. We will examine briefly direct combustion, pyrolysis, gasification, anaerobic digestion, and fermentation of biomass and some of the chemicals that can be produced by utilizing these techniques.

Direct combustion of biomass for cooking and warmth is perhaps as ancient as humankind itself. Until relatively recent times, humans depended upon the open flame for both of these, and in much of the world this is still true. During the eighteenth and nineteenth centuries stoves and furnaces were developed. These devices were generally crude and universally inefficient. Improvements were slow in coming until the latter half of this century. In the past two decades very significant progress has been made in producing more efficient equipment. Newly designed wood/biomass combustion stoves and furnaces burn much more efficiently and thus produce increased usable heat per unit mass consumed, and also reduce air pollution as a side benefit. Another development of great consequence uses a new system to cogenerate heat and electricity. It combines a bagasse combustion or gasification unit with a steam-injected gas turbine to greatly enhance the conversion of material that would otherwise be wasted into steam and electricity. This technique is already being employed at several locations by sugar mill operators.

Bagasse was long considered a nuisance to sugar production. It was disposed of any way possible, including dumping into the sea, but more recently it has been used to generate process steam and electricity. For example, the island of Mauritius produces 16 percent of its electricity by burning sugar cane bagasse, and in Hawaii the sugar industry now provides 10 percent of the state's electricity, saving the purchase of about three million barrels of oil annually. It is estimated that if the old inefficient bagasse burners found in most mills were replaced with the new gasification-turbine technology, the factories could boost their production of electricity more than twentyfold.

Gas turbines are typically found in jet aircraft engines, and are being used increasingly for generation of electricity. The steam-injected gas turbine captures the turbine's high-temperature exhaust, uses it to form steam, and injects the steam back into the combustion chamber with the gaseous fuel, thereby increasing both efficiency and power output. Biomass gasifiers used for other purposes will be discussed more thoroughly below. The combination of gasification and steam-injected gas turbine to cogenerate heat and electricity is amenable to almost any biomass feedstock. Finding better and more efficient ways to use biomass is critical for Third World countries at this very moment, and will become critical to the rest of the world as the inevitable fossil fuel weaning process begins.

Direct combustion randomly breaks down the structure of biomass to produce heat and light. As a form of energy, heat can be used directly for warmth or cooking, or it can be converted, for example, into steam which has many uses. However, direct combustion cannot be employed to convert biomass into alternate forms of energy such as liquid fuels. These can be obtained from biomass, however, through thermochemical conversion.

One might question the wisdom of searching for an alternate liquid fuel, for many feel hydrogen to be the ultimate fuel. The advantages of hydrogen have been recognized for a long time. It is the cleanest fuel of all, producing only water and traces of nitrogen oxides, which can be removed when it burns. It also has the highest energy-to-weight ratio of any fuel. Unfortunately, its energy-to-volume ratio is low, requiring an inordinately large fuel tank by today's transportation standards. It is difficult to store, tends to embrittle metal, and there are no natural sources of pure hydrogen. At present, there are no economical methods of producing it either, although several methods under study appear promising. Most of these cleave water molecules into hydrogen and oxygen. Electrolysis of water produces hydrogen and oxygen but requires more energy than is released when hydrogen is burned. Use of solar power to produce hydrogen is more promising. Professor Hi Ti Tien of Michigan State University has developed a solar cell that provides hydrogen from seawater. This photochemical cell utilizes a semiconductor system and appears to be the first instance in which hydrogen is produced directly and inexpensively from seawater using a cheap polycrystalline semiconductor material. Interestingly, hydrogen can also be made from biomass by first converting it to methanol followed by its catalytic disassociation into hydrogen and carbon monoxide.

The existing transportation infrastructure is almost entirely dependent on liquid fuels, with gasoline as the dominant automobile fuel, kerosene for jet aircraft, and diesel for most trucks, buses, and agricultural implements. No other proven source (compressed gases, hydrogen, or advanced batteries) comes close to the convenience of liquid fuels because they pack a large amount of easily usable energy into a very small volume. For example, using hydrogen as the fuel and an overly large fuel tank, today's average auto could go only about 70 miles without refueling. Biomass is the only renewable energy form capable of satisfying the need for liquid transportation fuels. Because of the many problems associated with hydrogen's use, storage, transport, and the cost of converting the present fuels' infrastructure, the transition to hydrogen is likely to be slow, even after techniques are available to produce it economically. Accordingly, biomass will find its appropriate role whether as a source of hydrogen, chemicals, fuels, or all of these.

Biomass can be converted to liquid fuels by several alternate procedures, but thermochemical conversion provides the only direct route. It uses high temperatures to convert biomass directly to energy or, as preferred, to a liquid or gaseous fuel. To produce high-value liquid and gaseous fuels, the conditions of the thermochemical process must be carefully controlled. When heated in a reduced oxygen atmosphere, the long polymer chains of cellulose, hemicellulose, and lignin in biomass are broken down into simpler molecules that are useful directly as gaseous fuels or can be converted to liquid fuels such as gasoline, diesel, and methanol. Two types of thermochemical conversion processes are used to produce fuels: pyrolysis and gasification.

Pyrolysis of biomass produces a biocrude, a petroleum-like material that can be upgraded to high-grade liquid fuels such as gasoline. Pyrolysis techniques are capable of using any biomass material, even including such materials as municipal solid waste. Biocrude-derived gasoline is chemically equivalent to petroleum-derived gasoline and can be used directly without engine modification.

Although oxygen-restricted heating of biomass has been practiced for decades to produce a few products like methanol (once called wood alcohol) and charcoal, processes that efficiently convert biomass into biocrude are very new, having been developed only since the early 1970s. The capability of upgrading these into gasoline is even newer. There are two steps involved: conversion of the biomass feedstock into a biocrude liquid, followed by upgrading the liquid through advanced catalytic processes to gasoline. The gasoline produced has an octane rating of 76, which is comparable to straight-run gasoline from petroleum. In the late 1970s and early 1980s, conversion yields of 20 percent of the dry biomass into biocrude were the norm. Today the yield is about 60 percent, which represents more than 70 percent of the energy content of the feedstock. Yields are still improving. Conversions of 60–80 percent will produce 75–100 gallons of gasoline per ton of dry biomass. Today's technology produces gasoline at $2–$3 per gallon. With anticipated process improvements the price should soon be less than $1 per gallon. At that price it will be competitive with petroleum as the price of crude oil rises above $20/bbl.

The amount and quality of biocrude obtained from pyrolysis is dependent on both feedstock and reaction conditions. For example, feedstocks with high lignin content produce more and better biocrude. This may foretell separating lignocellulosic materials into its components, fermenting the sugars, and pyrolyzing the lignins. The quality of biocrude liquids is also altered substantially by such things as pyrolysis temperature, heat-up rate, pressure, and catalysts. The two systems receiving the most attention are low- and high-pressure pyrolysis. Both methods are dependent on fast (very hot-short residence time) pyrolysis.

In the low-pressure method, the feedstock decomposes into vaporized small organic fragments that contain large amounts (up to 33 percent) of undesirable oxygen. The condensed vapors contain more than 70 percent of the energy in the feedstock and the procedure is inexpensive. However, the condensed material must be treated further to be very useful. One method is to pass the vapors over a zeolite catalyst which causes the vapor molecules to break apart, producing two fractions; one fraction contains gasoline-like material with octane numbers over 100 and the other is mostly tar. Work is underway to maximize the gasoline fraction at the expense of the tar.

The high-pressure pyrolysis method produces similar yields, but a more stable, lower oxygen (10 percent) content biocrude. It is more suited to normal refining techniques. In this method, feedstocks are mixed with recycled biocrude and heated at high pressure in the presence of a catalyst and a reducing gas. Although the procedure is more complicated and expensive, the improved product makes it worthwhile. Treating the biocrude produced by this method with hydrogen, in the presence of a catalyst, removes much of the oxygen and adds hydrogen to the reaction products, resulting in the production of a good-quality gasoline.

Tremendous technical advances have been made in pyrolysis during the past ten years and there is potential for much more. As the overall use of biomass becomes more sophisticated, there will be continued progress in both pyrolysis methodology and feedstock selection, yielding better fuels at lower costs. In addition to fuels, pyrolysis is capable of producing other useful products, some of which are even now commercial. Most are derived from high-lignin feedstock, are phenolic in nature, and are being used like phenol-formaldehyde thermosetting resins in the production of plywood, chipboard, and other lumber industry products.

The alternate thermochemical process to pyrolysis to produce fuels and chemical feedstocks is gasification. This method has actually been in use much longer than pyrolysis and differs from it primarily in temperature, reaction time and especially in that heat is supplied to the reactor indirectly or by burning a portion of the feedstock itself. The chief products of gasification are tar and synthesis gas (syngas), whereby the tar often serves as the source of heat for gasification. Syngas is a gaseous mixture composed mainly of carbon monoxide, hydrogen, carbon dioxide, and other gases such as methane and ethane. Depending on the feedstock and reaction conditions, other low-molecular-weight alkanes and olefins also can be produced.

As early as 1792, gasification of coal was used to produce gas which lighted houses and streets in England. Many cities of Europe and America soon followed the English example. This "coal gas," as it was often called, was used extensively until replaced by the more economical natural gas when an extensive distribution system was developed in the early decades of the twentieth century. However, gasification was again used extensively throughout Europe during World War II to produce fuels, both gaseous and liquid, when petroleum shortages occurred. Immediately after the war, interest in gasification all but disappeared and was not revived until the 1970s during periods of high energy prices and restricted fuel supplies. There is much renewed interest today because of the technical advances which make gasification of biomass a competitive source of energy and chemical feedstock.

The heating value of the syngas produced by gasification depends to a large degree on the source of the heat and oxygen to operate the system. Gasification can produce gas of low-BTU value (90–250 BTU per cubic foot), medium-BTU value (250–500 BTU per cubic foot), or high-BTU value (500–1,000 BTU per cubic foot). Low-value gas is produced if char and air are the sources of heat and oxygen. If an external source of heat is provided or pure oxygen used in conjunction with char, a medium-value gas is produced. In the absence of oxygen, a high-value gas can be produced if sufficient external heat is provided, but the process is not cost-effective using today's technology. Low-BTU syngas is best used as a boiler fuel, and can be utilized by boilers of any design. This, and the fact that there are skid-mounted gasification units already on the market, provides great flexibility and a large number of applications. Medium-BTU syngas possesses from one-third to one-half the energy content of natural gas. As such, it can be substituted for natural gas or petroleum as a fuel or can be catalytically converted into methanol or other alcohols, gasoline, diesel, jet fuel, or chemical feedstock. As such, medium-BTU syngas also has many diverse applications.

There are at least four processes that are capable of converting 90–95 percent of a feedstock to medium-BTU gases with energy efficiencies around 75 percent. They differ greatly in design; from an internal, fired-tube heat exchanger bundle, a directly heated, pressurized, fluid-bed gasifier, an indirectly heated, entrained-bed gasifier that uses hot sand as a heat source, to a directly heated, pressurized, fixed-bed downdraft gasifier. Designs are now being tailored to specific feedstocks or syngas use. Syngas purification techniques have been developed to remove trace quantities of ash and contaminating organic liquids. A catalytic process has been developed that destroys 99 percent of the tars contained in the gas.

Perhaps the greatest value of medium-BTU syngas is its ability to be converted into methanol. Syngas is reacted catalytically at elevated temperature and pressure to combine its carbon monoxide and hydrogen content, forming methanol. The reverse reaction can be used to produce hydrogen and carbon monoxide when desired. Methanol has great appeal as a fuel because it is very clean-burning. It can also be used as a starting material to produce oc-

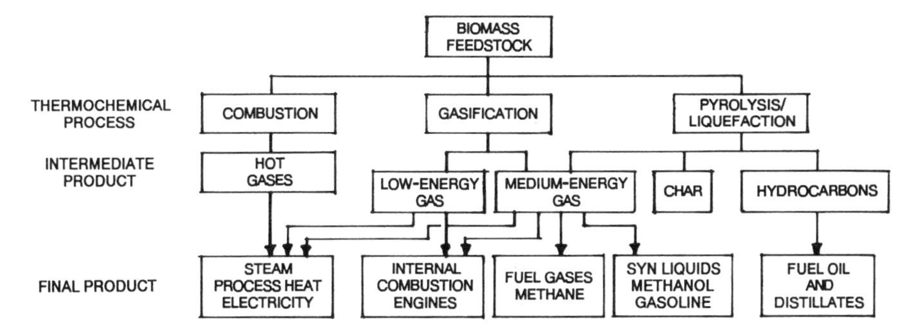

Fig. 6.1. Thermochemical conversions. By permission of G. F. Schiefelbein.

tane enhancers or reacted with zeolite catalysts to produce gasolines. The first commercial MTG (methanol-to-gasoline) plant was built in New Zealand using Mobil's process and zeolite catalyst, ZSM-5. In addition to gasoline, methanol can be used as a basic chemical feedstock that can provide most of the chemicals of today's commerce.

In summary, thermochemical conversion processes of biomass provide a broad spectrum of energy and chemical products. The same products are often provided by alternative procedures.

Biomass has distinct advantages over coal as a feedstock for thermochemical conversion processes. It has, for example, about a twofold higher content of volatile materials, its char is much more reactive than that of coal, it has a much lower ash content, and very much lower levels of sulfur. These characteristics allow biomass thermal systems to be designed that have relatively low capital investment requirements, making them much more economically attractive.

Another system for producing fuel which is becoming more attractive is anaerobic digestion, a special type of fermentation. This process, in which complex organic compounds are decomposed by anaerobic microorganisms, produces a "biogas" which is a mixture of methane and carbon dioxide. The heating value of this gas is about one-half that of natural gas because of the carbon dioxide component. As such it is classified as medium-BTU gas (500 BTU per cubic foot).

Biogas has been used as a fuel in various parts of the world for nearly a century. Manure is most widely used as a feedstock, with many small digesters providing fuel for cooking and lighting in rural India and China. Sanitary landfills operate as natural anaerobic digesters and some are being tapped for fuel. This uncontrolled operation is, of course, very inefficient. Recent efforts directed at reducing the amount of municipal waste have sparked renewed interest in biogas as a means of reducing the volume of municipal sludge and materials going into ever more expensive sanitary landfills. Properly managed, the biogas produced could reduce the volume of wastes and help offset the costs of waste treatment plants and sanitary landfills.

The process of anaerobic digestion is actually very complex and apparently carried out by a series of different microorganisms. At least three steps are involved: 1) breakdown of cellulose and other polymers enzymatically into simple sugars and other simple compounds; 2) bacterial digestion of these to form organic acids that are in turn reduced to smaller molecules of acetate, formate, hydrogen, and carbon dioxide; and 3) different bacteria called methanogens use these compounds to produce methane and carbon dioxide. The discovery that organisms that break down organic acids can be enhanced by light has dramatically changed reactor design and speeded up biogas production rates.

To make the production of biogas from lignocellulosic materials economical requires finding methods to increase the reactor solids loading rates, improving conversion efficiencies, and being able to control the process. Significant progress has been made in all three areas. Perhaps they can best be summarized by indicating the change in the cost of producing methane by this technique versus time. In 1980 the cost was $7.50 per million BTU methane. By 1987 the cost had dropped to $5.00 per million BTU methane and is projected to be $3.50 per million BTU methane by 1996. Current well head cost of natural gas averages about $2.00 per million BTU. It is projected to increase to around $3.00 per million BTU by the year 2000 and $3.75 per million BTU by 2005. Anaerobic digestion to produce biogas can make a very significant contribution to every urban area as well as to agricultural areas producing biomass.

Still another method of converting biomass to other substances utilizes fermentation. Virtually every culture throughout the ages and around the world has made alcoholic beverages and vinegar by this means. Biomass, however, was seldom if ever used because its cellulosic and hemicellulosic polymers must first be hydrolyzed to simple sugars before fermentation can proceed. The lignin fraction of biomass is not amenable to fermentation. Techniques for converting biomass into fermentation substrates and the fermentation process itself have advanced markedly during the past two decades. The stimulus and financial support for the research that made these advances possible were primarily born out of the search for alternative liquid fuels. Accordingly, the endpoint was ethanol rather than other chemicals. However, many of the procedures developed are just as applicable to production of other chemicals.

Ethanol has been the object of fuels research because it can be mixed with gasoline at concentrations of up to 20 percent with little or no adverse effect on current gasoline engines. It also has a very high octane value and consequently much of the premium unleaded gasoline sold is an ethanol-gasoline blend. These mixtures are cleaner-burning fuels than gasoline itself, especially at high altitudes, where the additional oxygen in the ethanol makes for cleaner burning. Many cities either already have required, or are considering requiring, the use of oxygenated fuels like "gasohol" to lower auto exhaust pollution levels. Currently, most fermentation ethanol is made from corn in the U.S. and from cane sugar products in other countries such as Brazil, the world's largest ethanol producer. Ethanol produced from corn is competitive in the U.S. only because of the current depressed farm economy, which means low corn prices, and the availability of federal tax credits that total $.60 per gallon of ethanol produced. These subsidies were due to expire in 1992, but have been extended, in part, through the year 2000. If alcohol is to be used as a fuel over an extended period it must use a feedstock, the price of which is not dependent on governmental subsidies. As indicated in chapter 5, according to economic studies, buffalo gourd even now would be a better raw material for producing alcohol than corn. The price of biomass feedstock is potentially lower than corn feedstock too, and also much more stable. The price of corn fluctuates dramatically because of many factors, e.g., weather, subsidies, governmental export policy, and its demand as a food or feed, while the price of biomass is not subject to such outside influences. During the past decade the cost of producing ethanol from biomass has decreased substantially from $4.00 per gallon in 1980 to $1.80 per gallon in 1987 to $1.35 per gallon as of this writing. The anticipated price by the turn of the century is expected to be about $0.60 per gallon, without tax credits, according to the U.S. Department of Energy's Solar Energy Research Institute. This price, if achieved, will make biomass-derived ethanol a very competitive fuel and chemical feedstock.

To be useful for fermentation, lignocellulosic material is usually first separated into its three main components—cellulose, hemicellulose, and lignin—by any one of several pretreatment steps. This can be accomplished, for example, by "steam explosion" of the biomass followed by water, very dilute acid or base, and in one system by a solvent extraction to remove the lignin. In any case the xylose, derived from hemicellulose, and sometimes the lignin, is separated from the crystalline cellulose. The xylose can then be fermented to ethanol or other chemicals and the lignin processed to produce either high-octane liquid fuels or functional chemicals. The crystalline cellulose remains behind as a solid after the pretreatment step and is hydrolyzed either enzymatically or by acid to produce glucose. In one scheme the glucose is then fermented to ethanol and combined with the ethanol from the xylose fermentation. The combined beers are subjected to distillation to produce ethanol.

The two most favored methods of hydrolyzing cellulose to glucose use either dilute acid or enzymes. Acid hydrolysis is fast, but conditions severe enough to break down cellulose to sugars also degrade the sugars into undesirable by-products, resulting in yield reduction. Systems have been developed which remove the sugars as they are formed, thus largely overcoming this problem. Enzymatic hydrolysis gives high yields of glucose, without undesirable by-products, but until recently it was slow. Cellulose enzyme hydrolysis technology is only about ten years old, and during that time most of its problems have been overcome. For example, a few years ago 60 to 70 percent conversion of cellulose to ethanol with ethanol concentrations of 2 percent were considered good. Today, conversion of over 90 percent is the rule with ethanol concentration approaching 5 percent. The amount of enzyme required has also been reduced by a factor of seven.

Ten years ago no methods were known for the fermentation of xylose to produce ethanol. Today, it is commonplace. In fact, it can be done directly by one fungal organism or by first treating xylose with xylose isomerase which converts it to xylulose. This sugar is easily and efficiently fermented by a yeast to ethanol. So many advances have been made in all aspects of converting biomass to ethanol that it now appears likely that simultaneous saccharification and fermentation can take the place of both cellulose and hemicellulose, thus producing high yields of ethanol, and leaving behind the lignin for other treatment. As the price of ethanol decreases, its use as a prime chemical feedstock will become common once again, as it was until the 1950s.

Fermentation used to be the prime method of producing many important chemicals. As a result of the recent advances in the direct treatment of biomass plus the significant improvements in fermentation technology, both have begun to be used commercially for chemical production. Brazil, for example, now uses ethanol to make a wide variety of chemicals, including ethylene. Fermentation and other biomass treatments are destined to become very important, if not dominant in chemical production. Biomass will likely replace oil as the chemical feedstock soon. Presently only 2 percent of the total oil consumption is used as chemical feedstock, but even this is an enormous quantity (about 350,000 bbls/day in the U.S.).

Prior to the post-World War II boom in the chemical industry and availability of low-cost petroleum and natural gas in the early 1950s, many of the most basic building blocks of the chemical industry were produced from sugar substrates by fermentation. These included such chemicals as ethanol, from which ethylene and butadiene were made, butanol, acetone, acetaldehyde, acetic acid, and at times chemicals like ethylene glycol and glycerol. Of the thousands of organic compounds now being produced, one hundred chemicals account for 99 percent of the total weight. Of these one hundred chemicals, 74 percent of them are produced from five primary compounds: ethylene, propylene, benzene, toluene, and xylene. Of these five, the first two are easily produced from the fermentation chemicals ethanol and isopropanol. The aromatics, benzene, toluene, and xylene, are produced by the thermal "cracking" of biocrudes from plants like *Grindelia camporum,* from selective treatment of lignin, or from syngas. Recent advances in metal-catalyzed reactions of methanol provide another route to the production of these aromatics. Recall that methanol is the primary product produced from syngas. Ethanol derivatives alone could account for a third of all organic chemicals, while propanol derivatives could account for another 10 percent. The list of organic compounds that can be produced from different methods of treating biomass as discussed above is enormous. In fact, there are probably none currently being made from petroleum feedstock that cannot be made from biomass. The key, of course, as to which is used is economics.

It is a curious fact that most of the technical breakthroughs which now make biomass an attractive alternative feedstock to oil came about in an effort to produce the lowest value product, fuel, rather than more value-added products such as chemicals. The research effort that brought about these positive results was largely, but not wholly, government-sponsored. It is particularly noteworthy that some of these positive results were even being commercialized when oil was still very inexpensive, i.e., less than $20 per barrel, and before the Persian Gulf crisis of 1990–91. For example, one potential ethanol-from-biomass producer was selling the isolated cellulose rather than converting it to ethanol, for the demand for cellulose was high and its direct sale more profitable than converting it to ethanol. The cellulose was converted to rayon and cellulose acetate. Also, steam-hydrolyzed lignocellulose fiber can be fed directly to cattle at a feed value equivalent to hay. Another example is alkaline hydrogen-peroxide-treated straw. This is as good as corn for animal feed and is being mass-produced today. Biomass-derived adhesives offer several important benefits to the wood products and construction industries. Their performance appears to equal that of conventional phenol-formaldehyde thermosetting resins at about half the raw material cost. Lignin, in various product forms, is being marketed in at least twenty-three use areas by six lignin producers. Its uses include such things as oil field and agricultural chemicals, asphalt extenders, carbon black, adhesives, engineering plastics, and specialty dispersants. Recent discoveries for treatment of lignocellulosic materials allow them to be brought into solution, molded, extruded, or cast into films. Depending on the treatment, thermoplastic or thermosetting properties may be imparted as well. Accordingly, foams, like polyurethane, can be made, fibers spun, and boards produced in any desired shape and with special surface characteristics. Such direct use of lignocellulosic materials will permit many new fabricated products that should be, comparatively speaking, very inexpensive. In addition to ethanol, most of these biomass-derived products are already being sold today even though oil is still relatively cheap. As oil prices rise, biomass-based products will become even more attractive. In 1987, James E. McNabb of Conoco, pointed out that OPEC was operating at 60 percent of capacity at the time, and would reach 80 percent of capacity early in the 1990s. Irrespective of political considerations, he feels market power will shift from consumer to supplier with a resulting trend toward raising prices again. As a consequence, he forecasted that although oil prices will remain in the low $20s per barrel until 1990, they will rise to the mid-$30s per barrel by 1995 and to $50 per barrel by the year 2000. If this is true, it will greatly favor the use of biomass as a feedstock for the chemical industry, and even make biomass-derived fuels attractive.

Why, one might ask, have we devoted so much attention to the utilization of biomass in a book concerning arid-lands agriculture? The answer is that growing biomass and applying the new technologies for its use serves the needs of arid lands and its people extremely well. The newly developed and emerging technologies now make biomass production in these regions a viable and economically attractive alternative. This is particularly true as one considers the long-term soil advantages of growing perennial versus annual crops.

It is true that arid regions will not normally produce as much biomass per unit area as non-arid regions, but the only reason for this is the lack of water. In fact, given adequate water, areas of high solar incidence and high temperatures will generally outproduce those with more temperate climates. For example, the San Joaquin and Imperial valleys of California produce about half of all the vegetables consumed in the United States. So stated, we do not expect the arid and semiarid regions of the world to become the prime biomass producing areas of the world because water is, and will remain, the limiting factor for the amount of biomass that can be produced.

However, by selecting the appropriate low-water-use crops it is possible to produce very significant quantities of biomass with limited amounts of water. Switching from today's crops, which usually require 6–10 acre-feet of irrigation water per year to those requiring only 2–3 acre-feet per year, will still produce large amounts of biomass in addition to primary product. This biomass, called bagasse, can be converted into other forms of useful matter. For example, *Grindelia camporum* will produce at least 5 tons per acre of biomass consuming only 2.5 acre-feet of applied water in areas receiving less than one foot of rainfall annually. After extraction of the resin, over 4 tons of bagasse remain for alternative uses, such as conversion into energy, fuels, or discrete chemicals.

Therefore, although water is the limiting resource in arid and semiarid lands, it need not impede the production of new crops which directly or indirectly produce biomass in good quantity to be used for alternate purposes as described earlier in this chapter.

Desertification is a serious problem worldwide. Denuding the land has fostered erosion as well as set the stage for diminished rainfall and encroachment of blowing sand on once productive soil. Biomass crops like most of those discussed in the previous chapter and all trees are perennials, and as such retard erosion. Some species have been shown to slow or stop the desertification process or even reverse it. Recall, too, that in most Third World arid regions, wood is still the chief source of fuel for cooking and heating in rural areas and charcoal in urban areas, creating a desperate need for reforestation.

Aside from the Persian Gulf countries, most arid nations are not blessed with oil deposits, and accordingly must import their liquid fuel, fertilizer, and chemical requirements, which they can ill afford. To pay for these and other imports these poorer countries have been forced to grow annual "cash" crops, which has resulted in the displacement of traditional intercropping or perennial crops, movement of people to the cities, and further degradation of the land.

By growing biomass crops and applying the new technologies described, these countries could ameliorate many of their land problems, provide for much of their fuel and chemical needs, increase employment, and lessen their foreign debt burdens. There are many potential plant species from which to choose the ones best-suited to any particular locale. Although only a few arid-adapted plant species have been mentioned in this book, there are many from which to choose. For example, there are at least seventy species of eucalyptus trees alone, most of which are rapid growers and arid-adapted. There are many acacias that would qualify too.

Even in wealthy countries like the United States, where much of the new technologies for biomass conversion has been developed, there are ample reasons to pursue biomass production.

The technology is still being improved. Moreover, it is available to be shared with the rest of the world. The time for government, agriculture, and associated industries to play a more significant role in producing chemicals, products, and fuels is at hand. Biomass cannot, in the foreseeable future, totally replace fossil fuels upon which the developed nations have become so dependent, but it can make a significant contribution. It can have an immediate impact on the more value-added markets such as chemicals, functional polymers, resins, and rubber to the financial benefit of the participating companies. It now appears that the production of fuels, liquid fuels in particular, can become a profitable venture even without governmental subsidies.

SELECTED INFORMATION SOURCES

Anaerobic Digestion Annual Report. 1989. FY 1988 SERI Report SP-231-3520, June.

Anderson, E. 1988. "Ethanol Producer Looks to Hardwoods as Raw Material." *Chemical and Engineering News,* January 4.

Anderson, E. 1989. "Brazil's Fuel Ethanol Program Comes Under Fire." *Chemical and Engineering News,* March 20.

Busche, R. M. 1988. "The Biomass Alternative." Tenth Symposium on Biotechnology for Fuels and Chemicals, Gatlinburg, Tenn., May.

The Economic Contribution of Lignins to Ethanol Production from Biomass. 1985. SERI Report TR-231-2488, May.

Ethanol From Biomass Annual Report. 1989. FY 1988 SERI Report SP-231-3521, June.

Five-Year Research Plan 1988–1992, Biofuels: Renewable Fuels for the Future. 1988. DOE Report CH10093-25, July.

Haggin, J. 1989. "Alternative Fuels to Petroleum Gain Increased Attention." *Chemical and Engineering News,* August 14.

Ng, T. K., R. M. Busche, C. C. McDonald, and R. W. F. Hardy. 1983. "Production of Feedstock Chemicals." *Science* 219(February 11):733–740.

Thermochemical Conversion Program Annual Meeting. 1988. SERI Report CP-231-3355, July.

Worthy, W. 1990. "Lignocelluloses Promise Improved Products for Materials Industries." *Chemical and Engineering News,* January 15.

SEVEN

Achieving the Promise

Homo sapiens is an endangered species. Not from the actions of nature or of another creature, but from our own acts. The development of the atomic bomb and nuclear energy provide the potential for self-destruction, but these do not necessarily represent the greatest threat today. From the point of view of supporting human life, the earth's physical condition is deteriorating at an alarming rate. Humanmade pollutants are fouling the atmosphere, leading to the potentially lethal effects of global warming, acid rain, smog, and stratospheric depletion of ozone. Forests are being destroyed, deserts are expanding, water shortages and contamination are mounting, and soils are eroding—all at unprecedented rates as the result of human activities. With the human population exploding and agricultural expansion declining, hunger, humankind's oldest enemy, is taking its toll. Each year there are 90 million more people to feed from 24 billion fewer tons of topsoil. Based on cold, hard data, the National Academy of Sciences recently warned President Bush that "global environmental change may well be the most pressing international issue of the next century," and "the future welfare of human society is . . . at risk."

Unlike the now-extinct thousands of species, human beings are endowed with intelligence and free will. Our survival depends to a large extent on how we choose to use these prerogatives. We may be very close to the point of no return, but many thoughtful people believe ultimate disaster is avoidable if the right choices are made *in time.* The purpose of this chapter is to examine the available choices and to indicate the promise the right choices offer with special regard for those that apply to arid lands.

Water: How to Stretch the Dwindling Resource

Water is essential for food production, industrial development, urban growth, and for advanced living standards everywhere. Incentives to use water more efficiently, to get the most out of every gallon used, are essential. Yet in most of the world the policies and laws governing water use discourage efficiency and in many cases actually encourage waste. Some examples of this misuse of a valuable resource are given in chapters 1 and 2. An economist for the World Resources Institute reported that government revenues from irrigation projects in six Asian countries averaged less than 10 percent of the actual cost of delivering the water. As a consequence, the governments were financially unable to properly maintain their projects and the farmers had virtually no incentive to conserve water.

As discussed in chapter 2, this problem is severe in the United States. To help settle the arid West, in 1902 Congress passed the Reclamation Act authorizing the Bureau of Reclamation to provide water under long-term contracts (usually 40 years) at greatly subsidized prices. Today, some 11 million acres, or a quarter of the West's irrigated land, get water from these federal projects. According to Sandra Postel of Worldwatch, the California farmers benefiting from the federal Central Valley Project have repaid only 5 percent of the actual cost in the last 40 years. This amounts to a total subsidy of over $930 million paid by the taxpayers.

In the San Joaquin Valley, nearly 20,000 farmers, some of whom own rich agribusinesses and control 10,000 acres or more, pay only $2.50 to $19.31 per acre-foot to irrigate crops with water from the federal Central Valley Project. Originally, subsidized water was intended for small family farms not exceeding 160 acres, but today owners of large farms use loopholes to get around the acreage limitaion for obtaining subsidized water. Urban residents in the six southern counties of California get most of their water from the

Metropolitan Water District at a cost of $233 per acre-foot. It is not surprising that there are proponents arguing for shifting water from agricultural use to urban use. They say the subsidized water going to farmers in the Central Valley has stimulated wasteful and inefficient agricultural practices.

About one-third of this highly subidized water is used to grow low-value crops such as grass and alfalfa to feed cattle and sheep. Irrigated pasture contributes little to the California economy (only $94 million of the state's gross product of $600 billion in 1986), but it consumed enough water to provide for the domestic needs of about 28 million Californians. Assemblyman Jim Costa defends the use of subidized water for alfalfa and pasture because it supports the state's dairy industry, which he claims accounts for $3.2 billion annually. Environmentalist Marc Reisner argues that support of a local dairy industry may have made sense at the turn of the century when the American West was isolated from the rest of the world by thousands of miles of ocean to the west and by hundreds of miles of desert to the east. The desire for self-sufficiency in food production was understandable then, but today with fast highways and refrigerated trucks and railroad cars, Californians could get their fresh meat and dairy products from states where irrigation water is not needed to grow grass and alfalfa.

California farmers grow other low-value crops that require enormous amounts of the government-subsidized water. Cotton—which other farmers are paid not to grow because of surpluses—consumes as much water as the cities of Los Angeles and San Francisco combined. And rice—which requires 80 inches of water per year—has been in surplus for 10 years and costs the U.S. over $1 billion in subsidies annually and contributes little to the economy of California. If these farmers would substitute appropriate arid-land crops, such as those described in chapters 4 nd 5, for just half of the grass, cotton, alfalfa, and rice they now grow, it would free up enough water to supply ten cities the size of Los Angeles. This would forestall the need for any new dams or aqueducts, and would provide adequate water for crops such as citrus fruit, grapes, tomatoes, artichokes, pears, peaches, almonds, and other high-value foods. In addition, there would be water available for rivers and wetlands to preserve what is left of the salmon and steelhead fishery, waterfowl, and other threatened wildlife.

California Assemblyman Phil Isenberg, the leader of a coalition of urban state legislators advocating changes in water policy, argues that there is no reason for irrigating anything in a state as dry as California. He and other critics of the allocation system claim that a shift of only 10 percent of water away from agriculture could provide for the urban needs of California for the next 20 to 30 years. They propose that agricultural irrigation districts should be allowed to sell some of their water to urban areas for a profit.

Professor Richard Howitt of the University of California at Davis conducted a study using a computer model to predict what would happen if water were permitted to move without state and local restrictions from farms to the cities. The conclusion from the study was that farmers could sell their water for more than they could make growing crops, and the cities could buy it for less than they would have to pay to develop new water projects. Under Howitt's model, only 8 percent of California's developed water was shifted from agricultural to urban use.

Considering urban water use in the arid southwest, Los Angeles, with its 12 million people, is probably the most prodigious urban user of water in the region. Annual average rainfall in southern California is 14 inches or less. The region is usually rainless from April through November and is at this writing into its fifth year of drought. But at least until very recently, the lawns remained green, swimming pools remained full, 8 million cars remained washed, and there were verdant cemeteries and countless golf courses. All this is possible because Los Angeles has taken the entire flow of the Owens River, a good share of the Colorado River and a third of the Feather River through an aqueduct 445 miles long. It is no wonder that this extravagant consumption of water in a desert region creates the notion that Los Angeles gobbles up half the water available to the whole state. But the truth is that Los Angeles is not the single biggest consumer of water in the state—it is not even close. In fact, all of urban southern California uses less than 10 percent of the state's water. Agriculture consumes by far the largest share, 83 percent of the 34.2 million acre-feet of the water provided by the state and federal systems.

California is not the only state in the arid west with problems resulting from poor management of water resources. In Nevada the environmentally important Stillwater wetland acreage has shrunk from 79,000 acres in the early 1900s to barely 5,000 acres today, much of it polluted with selenium, boron, and agricultural waste. The main cause: upstream water diversion to support an alfalfa economy contributing only a few million dollars per year to the state's economy. In Nevada, irrigation agriculture consumes 85 to 90 percent of all the available water, but contributes less than 5 percent to the state's economy. Conventional wisdom would have us believe that the big cities—Los Angeles, Tucson, Albuquerque, Denver, Phoenix—are the villains in the environmental decline of the West. While the cities are guilty to some extent where water is concerned, agriculture, or really western U.S. water policy, is the archvillain.

Is it fair to blame the farmers? According to Marc Reisner, author of *Cadillac Desert* and a recent article in *Greenpeace,* "farmers are merely pawns in an archaic system that has ceased to make much sense." It is unreasonable to expect them to refuse subsidized water that is too cheap to conserve. Under the existing system, to the farmer, saving water is more expensive than wasting it. For example, it would cost a farmer about $800 per acre to install a drip irrigation system, but he can flood irrigate for little more the $15 per acre. In an effort to bring some sense into western U.S. water policy, the Ford Foundation, with Reisner serving as consultant, sponsored an investigation and arrived at the following recommendations:

1. The Bureau of Reclamation should actively encourage transfer of water out of low-value agricultural uses by raising its absurdly low water rates.
2. In every western state water laws should be amended so that farmers who save water through conservation or by raising crops that require less water can retain their right to that water and should be permitted to sell it to make their effort and investment worthwhile.
3. In order to discourage frivolous water use and the raising of surplus crops, subsidized water used for these purposes should be subject to a hefty surtax.
4. Other western states should follow Oregon's example and amend their laws so that up to 25 percent of all salvaged or conserved water goes to fish, wildlife, and other environmental needs.
5. Cities seeking new water should first pay the farmers to install state-of-the-art irrigation technology so they can farm successfully with less water. This would make 10 to 30 percent of the water available to the cities.

Many water problems in other parts of the world are even more serious than those of the U.S. As pointed out in chapter 1, these problems are severe and worsening in North Africa and the Middle East. Egypt's 55 million people are almost completely dependent on water supplied by the Nile River. About 80 percent of the Nile flowage is from the Blue Nile with its headwaters in Ethiopia, and 20 percent from the White Nile which forms in Tanzania. Altogether nine countries draw water from the Nile basin and Egypt is the last in line. The two Niles merge in Sudan and according to a 1959 agreement, Egypt is entitled to 55.5 billion cubic meters (14,661.6 billion gallons) of Nile water per year. Even though Sudan may be willing to honor this agreement indefinitely, Ethiopia has development plans which could reduce the flow of the Blue Nile enough to cause serious consequences downstream.

Drought is another matter for concern. During the dry periods of the mid-1980s, the flow of the Nile into the impoundment at Aswan dropped well below Egypt's allotment, and by 1986 water storage was down to less than one-fifth of the reservoir's capacity. Heavy rains in 1988 brought the water level up close to the predrought level, but long-term records indicate a good possibility for low-flow periods at the beginning of each century. With Egypt's population growing at the rate of a million every nine months, and no additional water sources to draw on, the demands for water both for domestic use and food production are becoming critical. Egypt must learn to use water more efficiently and do a better job of managing its irrigation systems. The country imports one-half or more of its food, but still grows water-intensive cash crops such as cotton and sugar for export. Foreign exchange earnings are needed to offset Egypt's $44 billion external debt. If Egypt would utilize some of the freshwater-sparing techniques discussed in chapter 3 and learn to grow appropriate food and nonfood crops described in chapters 4 and 5, the nation could increase its food production and export industrial chemicals to improve foreign exchange earnings.

Water problems are particularly critical in Africa. Nowhere in the world is agricultural breakdown more serious than in the countries bordering the Sahara Desert on the south. Here, where the population growth is the highest of any continent, the combination of deforestation, drought, erosion, overgrazing, and civil strife have lowered per capita food production to an all-time low. The relatively few attempts at large-scale irrigation have for the most part been costly failures. Small-scale irrigation and watershed management projects that supplement traditional practices and familiar methods are proving useful in dealing with the water problems in this region. These techniques are presented in chapter 3, and appropriate crops are discussed in chapters 4 and 5.

The Middle Eastern countries in the Jordan River basin and in the Tigris-Euphrates watersheds are faced with similar problems. The farmers of Israel are probably the most water-efficient in the world, yet in the foreseeable future their water supplies could fall dangerously short of demand. The aquifer underlying the Gaza Strip supplies much of the water for northern and central Israel. Overpumping has caused seawater to enter this vital source. Jordan, with its population expanding at 3.6 percent per year, projects a 50 percent increase in water needs by the year 2005. Syria, Jordan's upstream neighbor, expects water shortages by 2000. In the Tigris-Euphrates watershed tensions are mounting. Both rivers rise in the mountains of eastern Turkey. The Euphrates flows through Syria and Iraq on its way to the Persian Gulf. The Tigris flows from Turkey directly through Iraq to the Persian Gulf. Turkey has an ambitious hydroelectric and irrigation project underway. When the Ataturk Dam is completed, it may hold back enough of the Euphrates flow to foil both Syrian and Iraqi plans to take more water for themselves.

India, with its 835 million people, is faced with worsening water shortages, resulting mainly from gross mismanagement of its land and water resources. During the last 20 years the government built over 1,500 large dams to provide hydroelectric power and irrigation water. Unfortunately, deforestation and overgrazing denuded the watersheds, causing floods that prevent percolation of rainfall into the ground to recharge aquifers. Erosion and salinization have degraded millions of acres of formerly productive land, and siltation has damaged many of the impoundments. Overdraft of groundwater in the western state of Gujarat has caused seawater to contaminate the aquifer. To the east, neighboring Bangladesh is at the mercy of India both for useful water and prevention of floods. Only recently this low-land nation suffered from catastrophic floods caused by deforestation in neighboring India.

China, with a population of 1.1 billion, also faces serious water constraints. More than 200 major cities lack sufficient water and they are challenging agriculture's claim on the scarce supplies. The shortage is at its worst in north China including the capital, Beijing, and the industrial port city of Tianjin. This region also includes farmland which produces a quarter of China's food. Industrial expansion, irrigation, and population growth have drained the region's meager water supplies to the point where rivers have dried up and overpumping of groundwater has emptied a third of the wells. Beijing's two main reservoirs are seriously depleted. Not only is water in critically short supply, but what they have is badly polluted. The volume of sewage generated in Beijing has increased twenty-sevenfold during the last 30 years and only 14 percent is treated. The rest is discharged directly into streams. China's officials and planners know that they must emphasize conservation and greater efficiency of water use, and they are giving agriculture lowest priority when allocating scarce water supplies.

In southeastern Australia salt buildup is diminishing the productivity of a million acres of prime farmland. This is a case of too much water in the wrong place. Here, water percolates downward to an impervious barrier of clay, forcing the water table to rise, bringing salts that have collected in the subsoil to the surface. When the water evaporates it leaves the surface salty, disrupts the structure of the soil, and hastens erosion. Traditional crops such as wheat and clover cannot tolerate the high salinity. The salty groundwater enters streams and pollutes drinking water. Salinity in the Murray River, which provides water for about half of Australia's irrigated farmland, has increased approximately 90 percent over the last 50 years.

This waterlogging and salinization were caused by two centuries of deforestation and European-style farming. The delicate water balance was upset. The deep-rooted trees and grasses native to the area took up rainfall and groundwater, minimizing runoff and deep penetration. The introduced European crops drew water only from the topsoil, allowing more to penetrate deeply into the subsoil. Australians in the states of Victoria, New South Wales, and South Australia have mounted a vigorous educational program to enlist grass-roots participation in a massive tree planting effort and in modifying irrigation and drainage methods to improve the situation. The newly planted trees are expected to draw water from the waterlogged subsoil to lower the water table. In addition, the trees will hold the soil in place, preventing erosion and runoff of salt into streams. Utilization of the salt-tolerant food and forage crops described in chapters 4 and 5 would be worth consideration by the Australians in their struggle to solve this problem.

People in all walks of life must encourage, even force, their governments to reexamine and take corrective measures as to water use and its cost. In the U.S., federal law must apply to ensure appropriate distribution, fairness of use and realistic cost, and to assure that special interest groups are not subsidized by the nation's taxpayers. The Ford Foundation's recommendations should be adopted by the federal government. In dealing with agricultural problems in most Third World countries, particularly in Africa, assistance policies should be rethought and localized, and small-scale projects should be employed in which the farmer himself (herself) has a large voice.

Global Warming—A Planetary Emergency

Today the news is full of stories on environmental tragedies, and there is a running debate among politicians, economists, and environmentalists on what should be done to solve the problems. Smog and other forms of air pollutuon have created health problems for the human population. The American Lung Association estimates that air pollution costs to the U.S. alone amount to $100 billion per year. Lakes, streams, forests, and crops are suffering from acid rain and the depletion of the ozone layer. Devastating oil spills occur all the time. But in the long run, the worst disaster of all may be global warming. The climatic changes, in terms of a human life span, may appear to be relatively slow and there are uncertainties involved, but uncertainty is no excuse for complacency.

The longer we wait the more serious the problem becomes. The changes in climate will be irreversible for any period of time important to us, our children, or our grandchildren. The critical need is to stabilize the greenhouse gas concentration in the atmosphere as soon as possible. The means to do this are known, but the task is monumental and global in scope. However, the results of the effort will affect virtually everyone in the world. Climatologists agree that we should begin by doing things that make sense in their own right, even if we did not have to face the threats of global warming. As Stephen Schneider says, "There are certain initiatives we can take that will buy us some planetary insurance and that will have other, ancillary benefits as well."

These initiatives include implementing an ambitious energy-efficiency plan, stimulating efforts to use renewable energy sources, banning chlorofluorocarbons (CFCs), stopping the destruction of forests, and starting a worldwide reforestation program. The industrial nations should assist Third World countries to achieve their development goals by using renewable rather than fossil fuels. The U.S. has a special responsibility to take the leadership with these initiatives. With only 5 percent of the world's population, Americans emit 23 percent of the world's carbon dioxide, an average of 18 tons per person per year. Japan, Germany, and Italy are more than twice as efficient as the U.S. in terms of the amount of energy used to make manufactured products. Increasing energy efficiency would not only reduce the hazards of climate change, but would also reduce acid rain, urban smog, and damage to crops and forests from air pollution. In addition, it would help the U.S. reduce its trade deficit (nearly 50 percent of the petroleum used is imported), enhance national security, and improve industrial competitiveness. The U.S. government uses high cost and loss of jobs as an excuse for delaying action to slow global warming, while other nations have been more analytical in their consideration of the problem. For example, a Canadian government-supported study found that although a 20 percent reduction in carbon emissions could cost the government $108 billion, it would also save the nation $192 billion through various benefits.

One way to implement an effective energy-efficiency plan would be to levy a "carbon tax" on fossil fuels. This would encourage individuals, companies, and governments to choose fuels based on their relative contributuon to global warming. Coal should be subject to the highest tax, oil next, then natural gas. Renewable energy sources, which do not contribute to global warming, should not be taxed at all. The United Kingdom's environmental secretary has proposed a carbon tax and the matter is under discussion internationally. For the U.S., this sort of tax could provide a threefold benefit: it would stimulate energy efficiency, reduce the trade deficit, and at the same time provide income to reduce the devestating national debt. The numbers involved are so enormous that even a modest per unit tax would yield a very substantial sum. For example, a tax of $1 per barrel on imported oil alone would yield over $6 billion per year (based on the 1989 import rate of 17,250,000 barrels per day).

To indicate the magnitude of the problem, today fossil fuels supply 78 percent of the world's energy. Oil provides 33 percent, coal 27 percent, and natural gas (mainly methane) 18 percent. The combustion of fossil fuels releases about 5.6 billion tons of carbon into the atmosphere annually. Fossil fuels vary widely in the amount of greenhouse gases they emit; natural gas is the cleanest, coal the worst. Per unit of energy, oil contains about 44 percent more carbon than natural gas and coal contains 75 percent more. Switching to natural gas has been suggested as an important strategy to deal with global warming. For the U.S., natural gas reserves are too limited for this to provide a long-term solution. Worldwide, natural gas is somewhat more plentiful and could provide some greenhouse benefit, but it is far from ideal. Since natural gas contains carbon it releases carbon dioxide when burned. Another disadvantage is that methane leaking from the distribution systems reduces the benefit of switching to natural gas because methane has a greenhouse effect 25 times more powerful than carbon dioxide. However, we could compensate for such losses by burning natural gas generated in landfills to supply useful energy rather than letting it escape into the atmosphere.

Coal is plentiful throughout the world and until the global warming crisis arose it was considered by many to be the most satisfactory substitute for petroleum. The quality and composition of coal varies widely from one location to another. The carbon content of coal averages around 73 percent, but ranges from less than 50 percent to nearly 90 percent. Virtually all coal contains some sulfur with low values around 0.5 percent and the upper contents greater than 6 percent. Coal also contains 1–2 percent nitrogen. These elements furnish the oxides that are converted in the atmosphere to sulfuric and nitric acids, the major components of acid rain. New power-plant technologies, developed jointly by the U.S. government and industry under the Clean Coal Demonstration Program, enacted in 1984, offer full-sized plant possibilities for reducing emissions of both sulfur dioxide and oxides of nitrogen, but no practical technology has been developed to "scrub" the carbon dioxide emission. While these advances are important in solving the acid rain problem, they do nothing to relieve the greenhouse effect of burning coal.

Improving the efficiency of fossil fuel use is clearly one of the most effective ways to reduce carbon emissions. Fuel not burned cannot produce carbon dioxide, smog, or acid rain, and it reduces dependence on imported oil. According to one energy expert, Amory Lovins, since the oil crisis in 1973, the U.S. has gained seven times as much energy from efficiency savings as from all increases in energy supply. Efficiency standards for automobiles, buildings, and appliances led to enormous energy savings and at the same time reduced pollution. For example, U.S. appliance standards alone are reported to have saved $28 billion worth of electricity and gas between 1986 and the end of the decade. And this kept over 340 million tons of carbon from entering the atmosphere. Reducing the highway speed limit to 55 miles per hour, increasing the miles per gallon only modestly, and use of the catalytic converter to reduce exhaust emissions provided similar spectacular benefits. Between the oil shocks of the 1970s and 1988, conservation enabled Americans to enjoy a 35 percent rise in the gross national product without increasing their energy consumption.

Unfortunately, the economic and environmental benefits stimulated by the OPEC embargoes diminished when cheaper gasoline prices returned. Gasoline consumption resumed its upward trend and car fuel efficiency progress slumped, especially in the U.S. Bigger, less efficient "muscle cars" came back in vogue. And the U.S., during the Reagan years, failed to formulate a meaningful energy plan. During his 1988 presidential campaign, George Bush promised, among other things, to be the "environmental president." As of the first quarter of 1990, his performance with regard to environmental matters has been disappointing to the concerned public. At the November 1989 Noordwijk high-level conference on atmospheric pollution and climatic change hosted by the Dutch government and attended by environmental ministers from over 70 nations, the world first officially recognized the need to stabilize (then reduce) emissions of greenhouse gases. Dr. Mostafa Tolba, director of the UN Environment Program, urged the delegates to recognize that "in the face of catastrophic possibilities, we cannot await empirical certainty. We know enough right now to begin action." Regretfully, with the U.S. leading the opposition, and Japan and the Soviet Union joining, all attempts to set any specific goals or timetables were blocked.

The Bush administration's position came under heavy attack, and a national opinion survey showed that American voters by a 3 to 1 margin wanted the U.S. to lead in fighting global warming. Intense political pressure apparently led President Bush to partially reverse his position in December 1989 during his summit meeting with Mikhail Gorbachev at Malta. But in February 1990, President Bush addressed the plenary session of the Intergovernmental Panel on Climate Change (IPCC), a high-level group of scientists, policy experts, and diplomats, with a highly disappointing message offering no real leadership and pledging only more study. The vast majority of nations, and the vast majority of Americans, seem unwilling to gamble with the earth's climate by postponing action. Hopefully, this will become clear even to the Bush administration before more valuable time is lost in dealing with this increasingly critical situation.

Renewable Alternatives to Fossil Fuels

Improvement in energy efficiency offers the best possibility for limiting carbon dioxide emissions in the near future. An energy efficiency improvement of only 3 percent worldwide would reduce carbon emissions by about 3 billion tons between 1990 and 2010. The other way to reduce carbon emissions is to use non-carbon-based energy sources. Arid lands offer some unique possibilities in this regard. Depletion of fossil fuel supplies will make this a necessity in the not-too-distant future, even if there was no global warming to deal with. Nuclear power does not produce greenhouse gases, but the unresolved problems of safety, mounting costs, difficulties in management, and disposal of the radioactive waste make it a poor option for solving global warming. The use of nuclear power today is displacing about 298 million tons of carbon emissions annually, or about 5 percent of the total. In the last 40 years or so, the U.S. has spent a trillion dollars trying to develop nuclear as a major source of energy, and as of 1991 it still provides no more energy than the burning of wood. The prospects for the contribution of nuclear energy to increase substantially are poor indeed. Alternative renewable energy sources that offer great promise for the near future include solar thermal, photovoltaic, wind power, water power, geothermal, and various biomass techniques. Arid lands often provide excellent opportunities to utilize these alternative energy sources.

Solar Thermal Power

In this solar power method, sunlight is focused from mirrored troughs into oil-filled tubes which convey heat to a turbine and generator to produce electrical power. During the last five years some 194 megawatts worth of generating capacity has been installed in southern California. These solar thermal systems convert up to 22 percent of the sunlight energy into electricity at a cost that is approaching 8 cents per kilowatt hour. By comparison, the recently commissioned, but still not operating, Seabrook nuclear plant in New Hampshire was expected to charge 12 cents per kilowatt hour. When the sun is not shining, natural gas serves as the back-up to run the turbines. These systems are practical for the sun belt regions and can compete economically with fossil-fuel-driven plants. In California, several solar thermal electric plants now sell electricity to utilities at competitive prices. Luz International, an Israeli builder of these solar thermal systems, is exploring the possibility of even larger plants in Nevada and Brazil. Considerable areas of land, as well as much sunshine, are required for large-scale solar thermal plants and deserts are best-suited to meet these requirements.

Solar Photovoltaics

Through the use of semiconductor materials, the sun's radiation is converted directly into electricity. Photovoltaic (PV) cells can be mounted on rooftops of buildings, recreational vehicles, and automobiles, and can provide energy to operate satellites, irrigation pumps, telephones, TV receivers, and billboards, or they can be installed in huge numbers as desert power plants. As Christopher Swan points out, "PV technology represents a quantum leap in simplifying the process of generating electricity." Solar energy, sunlight, is transformed into useful electricity requiring no fuel, no moving parts, no toxic emissions, no noise. This transformation takes place on an atomic scale within a thin wafer or coating. The U.S. government between 1958 and 1975 indirectly sponsored the industry by buying PVs for satellites under various NASA programs. Federal and state governments stimulated the industry's development with tax credits and research grants. Since 1975, research under the aegis of the Department of Energy, mainly through the Solar Energy Research Institute (SERI), has contributed important knowledge and assisted in the formation of many corporations. SERI research, in partnership with these corporations, resulted in marked declines in production costs and significant improvements in average cell efficiency. But the price of PVs is still not competitive in major utility markets without tax credits. And unfortunately, under the Reagan administration and again under George Bush, photovoltaic research financing has been slashed substantially, from a high of $150 million in 1981 to $35.5 million for fiscal 1989 and to a proposed budget of only $24 million for fiscal 1990. Thanks to a large extent to solar-conscious Assistant Secretary of Energy Mike Davis of the Bush administration, the solar budget for 1991 is $110 million and $146 million for 1992. Residential tax credit for solar heating was discontinued by the Reagan administration in 1985. This move brought about the loss of 35,000 jobs and wiped out over $500 million in annual sales of solar hot water systems.

The U.S. once had a dominant lead in solar research and development, but this has been lost thanks to low oil prices and unsympathetic government policy. At one time at least six domestic oil companies were engaged in research and development of solar energy, but by mid-1990 only two, Amoco and Mobil, were still involved. Arco, one of the leaders in the field, sold its solar subsidiary to the West German firm, Siemens AG. Now, Japanese and German government investments in photovoltaics research surpasses that of the U.S. and threatens the U.S. leadership position.

The lack of support by the U.S. government is regrettable, and should be reinstated, not only because it fails to provide protection for the environment, but also because it threatens jobs for American workers and foreign trade that was helping to solve the trade deficit problem. There are 250 U.S. PV manufacturers, distributors, and dealers. Currently, $150 million worth of solar panels are manufactured annually and about 65 percent of these are shipped overseas. Third World use of solar cells is expanding rapidly and is environmentally advantageous, offering an attractive market for U.S. firms if they are not undercut by government-subsidized Japanese and European competition.

Wind Power

This is one of the oldest of renewable energy sources, and during the last decade has become one of the most successful electricity-producing techniques. Today there are more than 20,000 wind turbines worldwide with capacity of some 1,600 megawatts. Most of the wind turbines have been installed in California and Denmark (Denmark leads the way in terms of quality of equipment), but wind power offers the potential for meeting a substantial portion of electricity needs in western U.S., northern Europe, India, North Africa, and the USSR. Wind power is one of the most economical of the emerging energy techniques. Modern turbines, in an appropriate location, can produce electricity at a cost of 6–8 cents per kilowatt-hour, with the potential for even lower costs in the near future. Wind power could provide more than 10 percent of the world's electrical power by the year 2030. Arid lands are often windy and provide good sites for wind turbines.

Water Power

As of 1986, hydroelectric power supplied 21 percent of the world's electricity, less than fossil fuels, but more than nuclear energy. The pros and cons of hydroelectric dams and irrigation projects have been discussed in earlier chapters, but in an appropriate site with proper construction and maintenance they can provide clean energy and useful water. The U.S. has the world's largest installed hydroelectric capacity, but the American dam-building era is about over. Industrial countries have already dammed their best sites, i.e., areas with steep, narrow gorges and adequate water flow. Most remaining possibilities have been reserved as parks or excluded from consideration because of their natural beauty or environmental importance. The potential for utilization of hydroelectric power is greatest in the developing countries of Asia, Latin America, and Africa. The World Bank indicates that 225,560 megawatts of large hydrocapacity will be added in developing countries by 1995. Over 40,000 megawatts capacity was installed between 1980 and 1985. More than half of this capacity is in Brazil, China, and India. The future for small hydroplants (with 15 megawatts capacity or less) appears favorable for Third World countries to provide power for isolated communities and agricultural processing plants far from established utility lines.

The water power of oceans is being investigated for generating electricity. Many arid lands lie next to oceans or seas. The objective is to capture the energy of tides, waves, and the temperature differential between surface and deep waters. Most of these efforts are still in the experimental stage, but a few commercial facilities have been developed. To generate electricity from the tides, a dam is placed across the mouth of a cove to form a pond. The rising tide is allowed to enter. After high tide, the gates are closed and the water is returned to the ocean through a turbine to generate electricity. The energy available from a tidal pond is proportional to the square of the tide's range. A minimal tidal range of 10 to 16 feet is required to be considered economically feasible. The upper end of the Bay of Fundy in Canada with tidal ranges of 30 to 40 feet is the world's greatest resource of tidal power. An 18-megawatt tidal plant is operating there now, and a 1,000-megawatt facility is being considered. Scientists are studying the environmental impact that such a large plant might have. The world's largest tidal plant is a 240-megawatt commercial unit at the La Rance estuary in northern France. The Soviets and the Chinese have built small tidal plants and are considering larger ones.

In 1986, the Norwegians, after years of experimentation, brought the world's first wave power plants on line. They have two prototype plants with a combined capacity of 0.85 megawatt. One of these plants, called an oscillating water column, produces electricity from both the rise and fall of individual waves, at a cost of only 4–6 cents per kilowatt-hour. The success of these prototypes has prompted Indonesia, Portugal, Puerto Rico, the United Kingdom, and Australia to follow suit. Japan operates several hundred small wave generators to power its navigational buoys and has tested larger turbines. Wave-generating technologies in the range of 1 to 20 kilowatts have been developed in the U.S., the Soviet Union, and Sweden.

With the oceans covering 70 percent of the earth's surface, they represent the largest solar collector on the planet. In the tropics, surface water temperatures are typically about 80 degrees F. and 39–40 degrees F. at 3,000–4,000-foot depths. This temperature differential can be used to generate electricity. In a closed-cycle system the warm surface water gives up its heat to a low-boiling fluid, such as ammonia or Freon, to power a turbogenerator. The vapor is condensed by the cold water from the depths. The U.S. Department of Energy is experimenting with an open-cycle system in which seawater is boiled in a vacuum chamber, producing both electricity and desalinated water. This development offers exciting possibilities for arid lands adjacent to oceans, e.g., Saudi Arabia, Baja California, Chile, and southern California. Another possibility for water power in both oceans and rivers are turbines that harness the energy of water currents. These turbines are somewhat like wind turbines and come in both horizontal-axis and vertical-axis varieties. Units based on these designs are considered economically feasible at flow speeds of 3 to 6 feet per second. Prototypes are being tested in the U.S., Canada, and Japan.

Geothermal Power

Geothermal power is generated from the earth's internal heat. Where geothermal capacity is available, electricity can be produced at the economical cost of 4–8 cents per kilowatt-hour. Worldwide, over 5000 megawatts of installed geothermal capacity is in operation today. The U.S. is producing about 2,000 megawatts now, but has the potential to increase this capacity severalfold. Pressurized hot water reserves are found in abundance in many regions throughout the world. New Zealand, Central America, the Soviet Union, and the Philippines could rely heavily on this renewable energy source for electrical power.

Biomass

This versatile energy source is capable of providing high-quality gaseous, liquid, and solid fuels in addition to electricity and chemicals. Wood and other vegetation and agricultural waste can be burned to produce heat and electricity or can be converted to alcohol and other chemicals as described in chapter 6. Burning biomass avoids increasing the greenhouse effect as long as the harvested materials are replaced with new growth. Biomass provides about 12 percent of the world's energy at the present time and could supply much more if it were more fully utilized. In order to make biomass fuels a more important energy supply, the world will have to grow biomass crops on marginal land and develop integrated cropping systems to allow the same land to produce food, chemicals, and fuel crops. These issues are addressed in previous chapters.

The Ultimate Fuel

Over the long term, hydrogen appears to be the fuel of choice. It is potentially plentiful, clean, and relatively inexpensive. It is nontoxic, noncarcinogenic, and poses no greater safety hazard than gasoline. When it burns, it simply combines with oxygen to form water, so it has no greenhouse effect and cannot deplete the ozone layer. Small traces of nitrogen oxides are formed when hydrogen is used in internal combustion engines, but even this can be eliminated by technology currently under development. This involves a membrane to separate nitrogen from the air before it enters the combustion chamber. Peter Hoffmann, editor of *The Hydrogen Letter,* points out, "You could burn the stuff indoors without a chimney." Energy experts around the world are showing vigorous interest in hydrogen, considered by many to be the environmentalist's dream come true.

Hydrogen can be produced by electrolysis of water. The main drawback is that more energy is required for a current to separate water into hydrogen and oxygen gases than is released when hydrogen is burned. The best way to produce hydrogen is with solar power, and advances in the efficiency of photovoltaic systems could make hydrogen preparation relatively inexpensive by the year 2000. Experts expect 12 to 18 percent efficient solar cells to be marketed by then. Joan Ogden and Robert Williams of Princeton University's Center for Energy and Environmental Studies suggest that arrays of photovoltaic cells could be located in the arid regions of southwestern U.S. to produce hydrogen gas. Studies have shown that high-pressure pipelines could then transport the hydrogen to homes and businesses around the country. It has also been shown that transporting energy in the form of hydrogen gas is less expensive than conducting an equivalent amount of energy as electricity through wire. Ogden and Williams have calculated that a collector field covering 24,000 square miles could produce hydrogen equivalent to the total oil consumption of the U.S. A collection field of this size would require 0.5 percent of the total U.S. land area, or about 7 percent of its desert area.

The fact that hydrogen is a gas at ordinary pressures and temperatures offers both disadvantages and advantages. Storage will probably be as a compressed liquid in tanks that resemble sophisticated thermos bottles. Storage tanks need to be much stronger than the sheet metal tanks used for gasoline or methanol. Leaks would be dangerous because of the potential for explosion, but hydrogen as the lightest of elements, dissipates rapidly instead of forming puddles or clouds like gasoline does. Europeans and Canadians are losing no time in researching the possibilities for hydrogen. The European Community and the dominion of Quebec along with a number of private companies and institutions have entered into a two-year effort to study the possibilities of making a fleet of buses powered by hydrogen produced by means of surplus Canadian hydroelectric power. German auto manufacturers BMW and Mercedes-Benz are investigating the possibilities of hydrogen-powered cars. Daimler-Benz has built a prototype car that runs on hydrogen energy and exemplifies one of the problems to overcome: hydrogen does not produce as much power as gasoline, so even with an overly large fuel tank it could go only about 70 miles without refueling. BMW is working with Bayernwerk AG to build a pilot plant to produce hydrogen using solar energy. The USSR is experimenting with hydrogen-powered airplanes.

Thus far the U.S. government has left the research and development of hydrogen largely to foreigners. The Department of Energy spends about $5 million a year on direct research, but the Environmental Protection Agency has emphasized methanol as the best clean fuel alternative to gasoline. Methanol could probably be available in quantity sooner and be slightly less costly than hydrogen, but although it is cleaner than gasoline, it still produces greenhouse gases and ozone, and is itself a toxic compound. Another drawback is that it would have to be mixed with about 15 percent gasoline to ensure that cars would start in cold weather. No matter what alternative fuel is chosen, massive retooling by the auto industry will be required, and building the infrastructure and distribution system for a new fuel will be a major undertaking. To do this for more than one alternative fuel would be unnecessarily expensive and awkward. It is important, therefore, that the research required to make the best choice of alternative fuel be conducted as competently and expeditiously as possible.

Government, Industry, and Academia— To the Rescue?

During the twentieth century U.S. agriculture has been enormously successful, and the system has been the envy of much of the rest of the world. In 1900 some 37 percent of the U.S. population depended on farming for a livelihood, and they produced food and fiber for themselves and the other 63 percent of the population. Today only 1 percent of the U.S. population produces the agricultural goods required for the entire population plus enough to supply a large export market. Clearly, modern conventional industrial farming has been highly successful in terms of total production and output per hour of human labor, but it is suffering from serious shortcomings when evaluated by other standards. There are concerns for human health and preservation of the environment resulting from the massive use of chemical fertilizers, pesticides, and herbicides, and the burning of huge amounts of fossil fuels in the heavy machinery involved. There are serious concerns over the sustainability of the resources that agriculture depends upon: erosion of topsoil, depletion of aquifers, salinization of irrigated land, compaction of soil, overgrazing of grasslands, loss of wet lands, destruction of forests, and contamination of groundwater.

U.S. agriculture is plagued with economic problems. During the 1980s about 25 percent of farms owned by individuals went out of business. This number would have been greater if the federal farm support programs had not escalated from $3.5 billion in 1978 to $26 billion in 1986. And even with the federal support programs, net farm income in terms of purchasing power has not risen significantly since 1970. Farm subsidies started in 1933 with Franklin Roosevelt's first "temporary" farm bill and every Congress since then has attempted to correct the problems caused by the preceding bill. This congressional tinkering has created a mind-boggling profusion of programs, each leading to new problems and more attempted corrections. Farmers left their farms and the farms grew in size. The number of farms shrank from 6.8 million in 1935 to about 2.2 million in 1987. Now about 73 percent of the farms are owned by nonresidents or hobbyists who earn most of their income outside the farm. About 27 percent of the farms produce nearly 90 percent of the total output and collect 87 percent of the federal farm subsidies. The large farms are sophisticated industrial operations. They are not self-sufficient like the family farms of the 1940s. They specialize in one or two commodities which will allow them to take full advantage of the government subsidies. As one critic put it, "They farm the government and milk the taxpayer." Rather than let the market determine prices and the acreage of each crop planted, Congress raised the farmer's income artificially and this has resulted in government ownership of a mountain of excess products. By and large, the survivors in American agriculture have learned to use the system to their own advantage and to the detriment of the national economy and the environment. This has caused gross overproduction and enormous surpluses of a few commodities and little or no production of others. It has contributed to the skyrocketing national debt and the multi-billion dollar trade deficit.

Contrary to what the American public appears to believe, today's farmers are not poor; their income and net worth are several times higher than the U.S. average. While most farmers remain hooked on conventional agricultural methods and federal aid, some more dedicated conservation-minded resident farmers are alarmed by the environmental and economic problems confronting the nation. They are concerned also for their own health and for the sustainability of their land. They are experimenting with what has come to be known as alternative agriculture. As early as 1976, the National Academy of Sciences (NAS), influenced by the Arab oil embargo of the early 1970s, published a landmark report emphasizing the need for developing renewable resources for industry. In addition, the NAS published many other reports on the development of specific new crops which could serve as renewable sources of rubber, oils, and other chemicals. Similar concerns were raised by the American Association for the Advancement of Science. In 1984, the Secretary of Agriculture appointed a task force made up of members from industry, government, and academia to study the challenges and opportunities facing American agriculture. Based upon a comprehensive analysis of the situation, in 1987 the New Farm and Forest Products Task Force concluded that diversification of agriculture and forestry must become a national priority. The main focus was on the development of new plant-derived farm and forestry products, including both new crops and new uses for existing crops. They concluded that this was required for American agriculture to meet the array of strategic industrial, economic, and social needs of the nation. The task force recommended the establishment of an independent organizational entity (somewhat like NIH) that would be dedicated to achievement of the diversification goal and provided with adequate resources to attain it.

During the latter months of 1988, at least five bills were introduced in the Congress that were apparently influenced by the task force recommendations. Unfortunately, despite effort and enthusiasm in both the House and Senate, all of the bills died before Christmas and the end of the Congressional session. In December 1988 the Association for the Advancement of Industrial Crops (AAIC) was formed. This organization recognized a serious problem with the inflexible agricultural bureaucracy, which had failed to adjust to changing times, and favors research on alternative industrial crops to reduce burdensome surpluses, to ease the troubles of the American farmer, and to reduce the national trade deficit. AAIC hopes to accomplish this through promoting research and publishing the results, raising the consciousness of public officials, by educating farmers and private individuals, and by pressing for legislation that will provide long-term assistance for developing new industrial crops including jojoba, guayule, cuphea, meadowfoam, lesquerella, kenaf, crambe, rabbitbrush, opuntias, and others. The first annual conference of the AAIC was held in October 1989 in Peoria, Illinois, cosponsored by the USDA/ARS Northern Regional Research Center. This was a truly international meeting. The keynote address was presented by Dr. Charles E. Hess, the USDA's assistant secretary of science and education. In his address entitled, "Commercialization and Utilization of New Industrial Crops," he pointed out that with President Bush's signing of P.L. 101-81 on August 14, 1989, the Agricultural Act of 1949 was amended to permit the planting of up to 20 percent of program acres to small oilseed and industrial crops without a loss of program base. In all, 92 papers were presented and abstracts published on all phases of new crops research, development, and utilization. The success of this conference should have contributed to the AAIC's objective of raising the consciousness of public officials and educating farmers and private individuals.

In September 1989, the National Research Council (NRC) of the National Academy of Sciences published its landmark report, *Alternative Agriculture,* after four years of study. The major conclusion of the NRC report: "A small number of farmers in most sections of U.S. agriculture currently use alternative farming systems, although components of alternate systems are used more widely. Farmers successfully adopting these systems generally derive significant sustained economic and environmental benefits. Wider adoption of proven alternative systems would result in greater economic benefits to farmers and environmental gains for the nation." The report also said that the "scientific knowledge, technology, and management skills necessary for widespread adoption of alternative agriculture are not widely available or well defined." Dr. Hess proclaimed the report "of unparalleled significance," but other individuals and organizations, including the American Farm Bureau Federation, Resources for the Future, and the Potash and Phosphate Institute, were critical, saying the report was biased and unscientific. Most of the chemical company spokesmen expressed a favorable attitude toward the NRC report. Will O. Carpenter, vice-president for technology at Monsanto Agricultural, said that "to protect soil and water, the move to sustainable methods will be a very positive thing," but cautioned that the NRC report is not "a set of prescriptions for all farmers" and "much more research needs to be done before all farmers can adopt alternative methods."

Chemical Warfare on the Farm

Initially, genetic research and biotechnology were expected to promote sustainable agriculture by providing increased yields, drought- and cold-tolerance, disease resistance, and even the hope to provide grain crops with the ability to fix nitrogen. The initial hope of these biochemical pursuits was to wean farmers away from the use of chemicals. Unfortunately, it appears now that this approach will only perpetuate the pesticide habit.

Researchers assumed at first that the desirable traits of food plants such as yield, plant size, drought-tolerance, etc. were monogenic, i.e., controlled by a single gene. It is now apparent that most of the genes for crop improvement are polygenic: traits are determined by numerous biochemical pathways; multiple genes govern their expression. Insurmountable difficulties have threatened hopes of miracle crop development, and companies that had invested millions of dollars in this research looked for alternatives. They sought traits that would be both financially rewarding and technically easy to transfer from one organism to another. Studies had shown that many weeds tolerated exposure to herbicides, and this ability to survive the toxin is governed by a single gene. Therefore, the industry's choice became herbicide resistance.

According to a recently published report, *Biotechnology's Bitter Harvest: Herbicide-Tolerant Crops and the Threat to Sustainable Agriculture,* twenty-seven corporations are working to develop almost all cereal and vegetable crops resistant to many of the major weed killers. Success with these endeavors will virtually force farmers to use more herbicides, more often, on more crops, resulting in more profits for the corporations involved and increasing costs to the farmer. The pesticide producers are not only usurping the well-being of the farmer, but that of the public as well. Millions of dollars worth of publicly funded research at state agricultural agencies and land grant universities are being devoted in support of corporate objectives.

Perhaps even more important is the likelihood that herbicide resistance will lead to environmental disaster. The adaptive nature of weeds to herbicides is well known. In ten years the number of species that have been documented to withstand several herbicides has increased from twelve to over fifty. Herbicde-resistant crops will surely lead to an increased use of weed killers. The weeds will adapt to the stress and new resistant weed species will develop. This will require the use of more and more chemicals to control them. The chemical manufacturers will prosper, but the farmer, the public, and the environment will suffer. Another ominous possibility is that the crop itself could become a pest. Clearly, the public and legislators need to be alerted to this unfortunate turn of events.

Legislative proposals are being made for the 1990 farm bill, and many, if enacted, will benefit alternative agriculture. Most of the proposals addressing agricultural commodity programs tend to give the farmers more flexibility in deciding what crops to plant. It is generally agreed that existing farm program rules are too restrictive and reward farmers for practices that are agronomically and environmentally unsound. The Bush administration bill, announced on February 6, 1990 by Agriculture Secretary Clayton K. Yeutter, would allow farmers to plant alternative crops without losing government benefits, and would use what was called "positive incentives" to reduce soil erosion, groundwater contamination, and other environmental problems. It also emphasized the need for more research and education. A coalition of environmental groups and several representatives have made a proposal that takes a more regulatory and activist approach. The Sustainable Agricultural Working Group has proposed legislation that would remove existing penalties on commodity programs and reward farmers for preserving and improving soil and water resources. Hopefully, the recommendation of the New Farm and Forest Products Task Force to establish an independent organizational unit dedicated to achieving diversification and providing adequate resources to attain it will not be forgotten.

The New Crop "Catch-22"

Development of a new crop is a high-risk, long-term proposition in which funding is hard to get before profit potential is demonstrated, and this cannot be done without funding for the initial research. A mechanism is needed to overcome this entry barrier so hindering to new crop development. Currently, neither the administration nor Congress adequately supports the high-risk venture of new crop development. Budgets are prepared by the USDA and the Office of Management and Budget. Congress then marks them up or down. Overcoming the entry barrier requires a multidisciplinary effort utilizing the talents of many specialists including those needed for germplasm collection, genetics, breeding, agronomy, physiology, pest management, crop modeling, processing, use, economics, and marketing. As Gary Jolliff has stated, "Congress should lead the way with legislative policies that would help overcome the multifaceted entry barrier—paving the way for aggressive, sustained, private sector investment by using, quite literally, the seed money of public funding." The USDA has the responsibility for germplasm enhancement, variety improvement, agronomic practices, and utilization of new crops. Industry has the responsibility to continue the research and move ahead with product development. The level of USDA involvement depends on Congressional support and appropriations. The problem is the staggering disproportion in expenditures between research and surplus commodity support programs. For example, the total USDA budget for research on industrial crops is only about $4 million annually out of a $1 billion total research budget, while in 1987 the corn-related support programs alone cost the U.S. government $12 billion. It is estimated that this amount spent on corn support programs for one year would be sufficient to support improvement research for 14 to 300 new crops for 20 years. And this does not take into consideration the economic opportunity losses from not growing profitable new crops on the millions of acres that may be producing subsidized surpluses or are idled by government payments.

However interesting and potentially profitable a venture into a biomass project may seem, it is justifiably a very serious and difficult decision to make in the affirmative. There is much corporate inertia to be overcome psychologically as well as practically. Excluding things like seed oil projects that fall into the normal pattern of agricultural endeavors, every program that harvests the whole plant is fraught with difficult and complex problems. Assume, for example, a company is interested in grindelia for its resins or guayule for its rubber because it fits the company's ongoing business and company objectives. As soon as the company does an economic analysis it realizes that it must "sell" every component of the entire plant, not just the resin or rubber, before the project has interesting financial return figures. This will likely take the company into sales areas in which it has neither experience nor company objectives. It requires development of a whole new business strategy as well as a new marketing effort. In the simplest case, but not necessarily the best, grindelia could be harvested for its resin and the bagasse used to cogenerate process energy and electricity, the latter being sold to the local utility. In the case of guayule, the interested company is most likely already in the rubber or resin business but not likely in both, or even if both, certainly not in the utility business, animal feed business, and/or pyrolysis, gasification, or fermentation businesses. To maximize profits, the entire plant must be utilized to its fullest potential. Aside from marketing considerations, the capital investment could be enormous, as would be the research effort to determine the optimum use of the entire plant. Even if we assume that the courageous and well-financed company made all the required research, engineering and capital investments to build the necessary facilities and marketing organizations to go forth with the project, it must still decide whether it also wants to become the farmer to grow the crop, with the attendent risks covered in a manner competitive with the federal goverment's policy, which presently offers the farmer a predetermined income for growing traditional crops. When all things are considered, it takes a well-financed company with bold management to venture forth into a new biomass-based business.

The most likely scenario to bring any of these crops to fruition, at least in the U.S., is a cooperative program in which the risks are shared between the government and private enterprise with government initially bearing the larger risk. The potential benefits to the nation are worthy of serious support: biomass/bagasse can provide a large, alternative renewable source of chemical feedstocks, liquid, and gaseous fuels, which can help both Third World countries and developed nations like the U.S. to be less dependent on imported fossil fuels. Biomass resources are generally dispersed and available throughout every nation, and can be grown in almost any climate, including arid regions. These crops are ecologically an improvement over most crops presently grown. Fuels derived from biomass are almost environmentally benign, being low in nitrogen and sulfur compounds. In addition, as stated before, biomass-derived fuels cannot overall add carbon dioxide to the atmosphere, but actually reduce it. They have few, if any, toxic effects on the air, land, or water. To produce biomass crops and to develop a biomass feedstocks industry can have a positive and widespread effect in the creation of jobs, the technology can be implemented without significant economic or social impact on surrounding communities, and there is already an existing infrastructure for this type of crop production in most countries. Overall, this type of crop for arid lands deserves serious and sustained support from both governmental bodies and private enterprise, and possibly, they will even be working together.

The National Plant Germplasm System

As mentioned above, the USDA has the responsibility for germplasm maintenance and enhancement, and this is carried out by the National Plant Germplasm System. A most important unit of this system is the National Seed Storage Laboratory (NSSL) in Fort Collins, Colorado where nearly a quarter of a million samples of crop seeds and their wild relatives are stored. This is the largest collection of its kind in the world. In addition, there are four regional plant introduction stations: Pullman, Washinton; Ames, Iowa; Experiment, Georgia; and Geneva, New York. What is the role and significance of this collection to American agriculture? To answer this question, we must consider the following: between the 1930s and 1970 acreage under cultivation in the U.S. fell by 15 percent, yet during this period total crop yields doubled. The use of fertilizers and irrigation contributed to this progress, but crop geneticists estimate that about half of the yield increase was due to the continued replacement of existing crops with improved varieties. Therefore, plant genetic material, germplasm, was the single biggest contributor to the growth of American agriculture during those years.

Today, plant germplasm is becoming more important than ever. Considering the urgent need to reduce the use of chemical pesticides, herbicides, and fertilizers, it is important to use genetic resources to breed new varieties better able to resist pests, diseases, and weed competition. Collections such as the NSSL's are becoming increasingly valuable because many of the samples are difficult or impossible to replace. Another reason for the increased importance of germplasm is the potential of biotechnology, which can create new varieties by the insertion of a single gene of interest into an individual plant. But biotechnology can't create new genes. The genes can only be found in nature. This is one reason why the destruction of rainforests is so disastrous because these regions represent the world's richest source of germplasm and species, both plant and animal, most of which have not been scientifically characterized. With the destruction of the rainforest, this treasure is lost forever. Now, in addition to pests, disease, water shortages, salinization, and loss of topsoil, agriculture is faced with the possibilities of a more exotic threat—global warming. While it is difficult to predict the timing, magnitude, or specific effects of global warming on agriculture, the only way to meet these uncertainties is through the maintenance of genetic diversity with a healthy seed bank.

Unfortunately, our seed bank, the NSSL, is not in a good state of health. At the end of 1988, fully 45 percent of the samples in the laboratory contained so few seeds that officials could not risk removing any more for tests to determine if they were still alive. In the past 5 years, only 36 percent of the collection had been tested for viability and of those tested one in four was in need of regeneration. The NSSL lacks the staff, the money, and the facilities to carry out its assignment of maintaining and evaluating its collection. The plight of the germplasm system derives fundamentally from neglect by the USDA, but the blame should be shared with Congress. The NSSL's budget for 1988 was $2 million, and this was twice the funding it had in 1987. The increase has enabled staff to be augmented and vital seed data to be transferred to computer, but this is only a small start toward solving a very big problem. Also, the U.S. Gerplasm System has no meaningful program in place to preserve any of the germplasm collected for new crop development. The National Plant Germplasm System seems not to be concerned at all. When one reads that Congress gave TV tycoon Ted Turner $5 million to support his 1990 Seattle sports spectacular, it would appear that Congressional priorities are badly distorted.

Because of a gross lack of information and understanding, the deterioration of the germplasm system fails to trigger a public outcry, such as the one precipitated by the use of Alar on apples. Some members of Congress do understand the importance of plant germplasm, but a major problem is the lack of a constituency. Representative George E. Brown (D-Calif.) and his staff on the House Agricultural Committee were working on a bill that would give the germplasm system its own budget and a specific authorization from Congress. The germplasm system would have a specific mandate, be required to report to Congress, and develop a comprehensive long-range plan. A germplasm advisory body of top scientists and policymakers would be established. This bill was supposed to be ready for introduction in the House in the spring of 1990. For the future of agriculture, it is important that initiatives such as this be adopted as soon as possible.

Two Examples of New Crop Development

The history of soybean production in the U.S. from a minor forage crop to the multibillion dollar crop of today provides an interesting example of new crop development. The soybean was introduced in North America over 200 years ago. Some authorities credit this to Samuel Bowen in 1765, while others claim that Benjamin Franklin sent America's first soybeans home in 1770 while he was a colonial agent in England. At any rate, little was done to develop the potential of this plant for over a hundred years, until acute shortages of vegetable oil during World Wars I and II sent soybean prices skyrocketing. In spite of the fact that soybean varieties available at that time were low-yielding, the high price suddenly made the soybean profitable. This crisis-born profit potential, plus the strong influence of W. J. Morse, director of soybean research in the USDA, provided the incentive to break the new crop entry barrier and the commercialization process was initiated. Publicly funded support was invested in soybean research, resulting in a threefold increase in crop yield between 1920 and 1985. Acreage planted in soybeans expanded rapidly to about 40 million acres and the crop became a multibillion dollar contributor to the nation's economy. Now this "miracle crop" is in danger of becoming a surplus commodity because of domestic overproduction and foreign competition.

The history of the Canadian rapeseed (canola) industry presents some similarities to that of the soybean in the U.S., plus some insightful differences. Recognizing the need for an alternative crop, the Saskatchewan Wheat Board selected rapeseed as a candidate and Canada committed about $40 million (Canadian) over a fourteen-year period to rapeseed breeding and genetic research for yield improvement and development of new varieties for edible oil production. They gave their product the more euphemistic name of canola. This edible oil has the advantage of containing the lowest amount of saturated fat (6 percent) of any vegetable oil and almost no erucic acid. This Canadian enterprise shows how a focused investment strategy applied to a carefully chosen candidate crop can overcome the entry barrier in a reasonably short time. One analysis indicated that the canola development provided a 101 percent annual net return to society, with benefits to both producers and consumers. Today canola is a major export earner for Canada.

The USDA has had a clear mandate for new crop development for a century, but it has not provided what the nation needs, probably because it does not have the institutional structure required. Institutional models for the kind of multidisciplinary effort needed do exist and have received monetary support from the U.S. The International Rice Research Institute in the Philippines, the International Potato Center in Peru, and the Centro Internacional de Mejoramiento de Maiz y Trigo (CIMMYT) in Mexico are examples of organizations that have been successful in developing new crops and improving established crops. The Canadian system worked well at least for the development of canola. In contrast to the slow commercialization of the soybean in the U.S., the successful commercialization of rapeseed in Canada in a minimum time demonstrates the effectiveness of a concentrated and cooperative research effort.

What needs to be done to hasten the development of the many promising arid-land candidates described in chapters 3 and 4? As Dr. Gary Jolliff stated in his presentation to the Senate Committee on Agriculture, Nutrition and Forestry Subcommittee on Agricultural Research and General Legislation in Washington, D.C. on July 27, 1989, "The high cost of chronic agricultural surpluses and acreage diversion programs justifies immediate national attention to strategic planning and development of new crops. The need to diversify the crop plant base from which the U.S. generates its annually renewable wealth is urgent." He went on to say, "The primary recommendation to the Secretary, United States Department of Agriculture, from the New Farm and Forest Products Task Force was the establishment of a Foundation for New Farm and Forest Products. The Task Force considered the establishment of an *independent* organizational entity, dedicated to the achievement of agricultural diversification to be absolutely vital."

At least a few members of Congress do think U.S. agriculture could use some institutional innovation. California Representative George E. Brown has pointed out that public interest groups have been calling for research in the areas of food, worker and environmental safety, and has published a paper entitled, "The Critical Challenges Facing the Structure and Function of Agricultural Research." Unfortunately, the general public is not well informed on agricultural matters in general, and on arid-land needs in particular. Powerful special interest groups are active in maintaining the status quo. It is vital that the American citizenry become more familiar with these issues, and that they express their desires forcefully to their legislators, local, state, and federal.

The Real Bottom Line

Judging from the problems discussed herein, it is apparent that the basic biological systems of the world are in trouble, but the economic indicators claim the world is prospering. Global economic output expanded by more than 20 percent during the 1980s, millions of new jobs were added, and international trade increased. The gross national product (GNP) for most countries, including the U.S., showed healthy gains. How can the economic indicators be so positive when the basic biological indicators are so negative? The answer is that the economic indicators are flawed. The classical way to calculate GNP includes depreciation of plant and equipment, but does not take into consideration the depreciation of natural capital such as nonrenewable resources including oil, forests, mined water, topsoil, farmland lost to urbanization and highways, etc. This flaw in the current GNP accounting can result in a misleading sense of economic health. Economist Robert Repetto and his colleagues at the World Resources Institute point out that the conventional system not only overstates progress, it may indicate progress when there is actually decline. For example, as Lester Brown of the Worldwatch Institute stated, "If adjustment were made for all grain produced with the unsustainable use of land and water, it would show a grain output well below consumption and provide a much bleaker sense of global food security." This concept has been considered in depth in a recent article by Sandra Postel in which she shows that a major restructuring of economic rules and practices is essential for environmental sustainability and for human survival.

Taking advantage of the possibilities to increase agricultural productivity and sustainability by utilizing the new crops and methods outlined above can improve the current precarious situation. But long term, even complete success in this regard will not solve humanity's most deepening predicament. Unless it is brought under control, population growth will overwhelm the earth's life support systems. The Club of Earth, a group of scientists whose members all belong to both the U.S. National Academy of Sciences and the American Academy of Arts and Sciences, stated it well in their release in September 1988, when they said in part:

> Arresting global population growth should be second in importance only to avoiding nuclear war on humanity's agenda. Overpopulation and rapid population growth are intimately connected with most aspects of the current human predicament, including rapid depletion of nonrenewable resources, deterioration of the environment (including rapid climate change), and increasing international tensions.

Therefore, an important prerequisite for achieving the promise is to bring population growth under control. Also, we must remind ourselves that an uninformed, undisciplined, nonparticipating populace can never expect to make a democracy function well. Complacency and ignorance on the part of the people make it possible for special interest groups to gain control of the government for their own sefish benefit. If democracies (including the U.S.) are going to survive and to function effectively for the people, then the people are going to have to be well-informed and participate effectively in their governments.

Conclusion

As we are writing these final lines on achieving the promise, the U.S. and the world are observing Earth Day 1990 on the twentieth anniversary of the first Earth Day. While it is encouraging to experience the enormous attention and publicity the media and celebrities are devoting to this event, there are two important aspects of the environmental crisis which receive little or no attention: one is that Earth Day is really a misnomer. What humankind is doing to damage the environment is not going to destroy the earth. The earth will survive even if the human life-support systems are completely destroyed. It is the survival of humankind that is threatened, not the survival of the planet. The second aspect completely ignored by the media is that concern for the environment did not begin with Earth Day 1970. It began many generations ago by sportspeople and naturalists such as outstanding conservationists John Lacey, John Muir, George Bird Grinnell, Theodore Roosevelt, Aldo Leopold, "Ding" Darling, and Paul R. Ehrlich, to mention a few. If their recommendations had been publicized and heeded on a global basis, there probably would have been little need for an Earth Day 1990.

The hour is late. But Nature is resilient and the promise is great. Intelligent *Homo sapiens* should be capable of achieving it. Science and technology along with a bountiful nature offer the wherewithal to deal with the problems, but education and obedience to the Ten Commandments and the Golden Rule are also required to improve our stewardship of earth and to solve humankind's problems.

SELECTED INFORMATION SOURCES

Balzhiser, R. E., and K. E. Yeager. 1987. "Coal-Fired Power Plants of the Future." *Scientific American* 257(3):100–107.

Bjerklie, D. 1990. "Greenhouse Initiatives." *Technology Review* 93(3):21–22.

Brown, G. E. 1989. "The Critical Challenges Facing the Structure and Function of Agricultural Research." *Journal of Production Agriculture* 2:98–102.

Brown, L. 1990. "The Illusion of Progress." In Lester R. Brown et al. eds., *State of the World 1990*, pp. 3–16. New York: W. W. Norton.

Dillingham, S. 1989. "Solar Industry Sees a Sunny Future." *Insight* 5(19):44–45.

Ehrlich, P. R., and A. H. Ehrlich. 1990. "The Population Explosion." *The Americus Journal* 12(1):22–29.

Ellis, W. S., and O. C. Turnley. 1990. "The Aral—A Soviet Sea Lies Dying." *National Geographic* 177(2):73–93.

Flavin, C. 1990. "Slowing Global Warming." In Lester R. Brown et al., eds., *State of the World 1990*, pp. 17–38. New York: W. W. Norton.

Hansen, P. 1989. "Energy Use and the Greenhouse Future." *Outdoor America* 54(3):21–24.

Hileman, B. 1990. "Alternative Agriculture." *Chemical and Engineering News* 68(10):26–40.

Houghton, R. A., and G. M. Woodwell. 1989. "Global Climatic Change." *Scientific American* 260(4):36–44.

Holzman, D. 1990. "A Lighter Side to Alternative Fuels." *Insight* 6(5):48–49.

Jolliff, G. D. 1989. "Strategic Planning for New Crop Development." *Journal of Production Agriculture* 2(1):6–13.

Lepkowski, W. 1989. "Farmers Urged to Adopt Alternative Agriculture." *Chemical and Engineering News* 67(37):5–6.

Lochhead, C. 1987. "The Farm." *Insight* 3(December 7):8–30.

McCoy-Thompson, M. 1989. "Saline Solution." *Worldwatch* 2(4):5–6.

Mohnen, V. A. 1988. "The Challenge of Acid Rain." *Scientific American* 259(2):30–38.

Peck, L. 1990. "Here Comes the Sun." *The Americus Journal* 12(2):27–32.

Postel, S. 1988. "North China Exceeds Its Water Budget." *WorldWatch* 1(5):40–41.

Postel, S. 1989. "Land's End." *Worldwatch* 2(3):12–20.

Postel, S. 1989. "Trouble on Tap." *Worldwatch* 2(5):12–20.

Postel, S. 1990. "Toward A New 'Eco-'Nomics." *Worldwatch* 3(5):21–28.

Press Briefing. 1990. "U.S. Rice Policy Harmful to Thailand." *Insight* 6(14):38.

Raeburn, P. 1989–1990. "Seeds of Despair." *Issues in Science and Technology* 6(2):71–76.

Ragan, M. L. 1991. "Amid Big Water Fight, California Is Still Dry." *Insight* 7(2):20–22.

Reisner, M. 1989. "The Emerald Desert." *Greenpeace* 14(4):6–10.

Romig, D. 1990. "New Routes in Conventional Agriculture—The Promises and Pitfalls of Herbicide-Resistant Crops." *The Land Report* 39:16–19.

Rosenfeld, A. H., and D. Hafemeister. 1988. "Energy-Efficient Buildings." *Scientific American* 258(4):78–85.

Schneider, S. H. 1988. "Doing Something About the Weather." *World Monitor* 1(3):28–37.

Schneider, S. H. 1989. *Global Warming: Are We Entering the Greenhouse Century?* San Francisco: Sierra Club.

Shea, C. P. 1988. *Renewable Energy: Today's Contribution, Tomorrow's Promise.* Paper 81. Washington, D.C.: Worldwatch Institute.

Stein, J. 1990. "Hydrogen: Clean, Safe and Inexhaustible." *The Americus Journal* 12(2):33–36.

Swan, C. C. 1986. *Suncell—Energy, Economy & Photovoltaics.* San Francisco: Sierra Club.

Udall, J. R. 1989. "Climate Shock, Turning Down the Heat." *Sierra* 74(4):26–33.

Wiley, K. 1990. "Untying the Western Water Knot." *The Nature Conservancy Magazine* 40(2):5–13.

Glossary

Accessions collections recorded in order of acquisition.

Acid hydrolysis the treatment of cellulosic materials using acid solutions to break down the cellulose to simple sugars.

Acre-foot 325,851 U.S. gallons: 12 inches of water covering one entire acre, enough to satisfy the needs of a family of four for two years.

Alcohols the family name of a group of organic compounds that vary in chain length and are composed of a hydrocarbon plus one or more hydroxyl groups, e.g., the straight chain series which includes methanol and ethanol. If the hydrocarbon is aromatic, the compounds are called phenols.

Alkane a saturated open-chain hydrocarbon.

Allergen a substance that induces allergy.

Anaerobic digestion degradation of organic matter by microbes in the absence of air (oxygen) to produce methane and carbon dioxide (biogas).

Antioxidant chemical which protects against oxidative degradation.

Aquifer an undergroud layer of porous rock, sand, etc. containing water.

Bagasse plant residue left after a product(s) has been extracted.

Biocrude a term used to describe a range of liquid products usually obtained from pyrolysis and direct liquefaction of lignocellulosics.

Biofuel fuel produced from biological materials.

Biogas a gaseous mixture of carbon dioxide and methane produced by the anaerobic digestion of organic matter.

Biomass organic, nonfossil material of biological origin constituting a renewable energy resource.

Bole the trunk of a tree, usually devoid of branches.

British Thermal Unit (BTU) the amount of heat required to raise the temperature of 1 pound of water 1 Fahrenheit degree under one stated condition of pressure and temperature; 1 MBTU equals 1 million BTU.

Caliche a crust of calcium carbonate that forms on the stony soil of arid regions.

Carbohydrate organic compounds made up of carbon, hydrogen, and oxygen which includes starches and sugars.

Catalyst any substance that facilitates the occurrence of a chemical reaction but does not itself undergo permanent change. In the presence of the appropriate catalyst, reactions that are slow to reach equilibrium are facilitated.

Cellulose the carbohydrate that is the principal constituent of wood and forms the structural framework of the wood cells.

Char a porous, solid carbonaceous residue resulting from the pyrolysis or incomplete combustion of organic material. If produced from coal, it is called coke; if produced from wood or bone, it is called charcoal. It is closer to pure carbon than the coal, wood, or bone from which it is produced.

Cofiring the use of a mixture of two fuels within the same combustion chamber; e.g., wood and coal.

Cogeneration the technology of producing electric energy and another form of useful energy (usually thermal) for industrial, commercial, or domestic heating or cooling purposes through the sequential use of the energy source, thus gaining in first-law efficiency.

Combustion an exothermic chemical reaction of a fuel with oxygen, often intended for the direct production of heat.

Coppice system a crop of trees that has been regenerated from the shoots of stumps and roots left after harvest.

Coproducts the resulting substances and materials that accompany the production of a primary product.

Cracking a reduction of molecular weight by breaking bonds, which may be done by thermal, catalytic, or hydrocracking. Heavy hydrocarbons, such as fuel oils, are broken up into lighter hydrocarbons, such as gasoline.

Dehiscence an act or instance of splitting open of seedpods to discharge contents at maturity, sometimes with great force.

Energy crops lignocellulosic feedstocks including herbaceous plants and short rotation woody trees grown under specialized conditions to maximize their suitability for energy production.

Enzymatic hydrolysis use of an enzyme to promote the conversion, by reaction with water, of a complex substance into two or more smaller molecules.

Enzymes a class of proteins that catalyze biochemical reactions.

Ethanol (ethyl alcohol, grain alcohol) can be produced chemically from ethylene or biologically from the fermentation of various carbohydrates found in agricultural crops and residues or from wood.

Feedstock any material used as a fuel directly or converted to another form of fuel or other products such as chemicals.

Fermentation decomposition of organic compounds, by microorganisms, to fuels and chemicals such as alcohols, acids, and energy-rich gases.

Fossil fuel organic fuel formed from the remains of plants or animals within or beneath the earth's crust; e.g., coal, petroleum, and natural gas.

Fuel pathway the steps in a process to biologically, thermally, or chemically produce renewable liquid and gaseous fuels from biomass, including municipal waste.

Gasification any chemical or heat process used to convert a feedstock to a gaseous fuel.

Gasifier a system that converts solid fuel to gas. Generally refers to thermochemical processes. Major types are designated as moving bed, fixed bed, entrained bed, and fluidized bed.

Genetic engineering the alteration of cellular function by introducing new genetic material into a cell.

Glucose a simple sugar containing six carbon atoms, six oxygen atoms, and twelve hydrogen atoms. A sweet colorless sugar that is the most common sugar in nature and the primary component of starch and cellulose. The sugar most commonly fermented by yeast to produce ethyl alcohol.

Graywater nonpotable irrigation water reclaimed from household and/or industrial uses, excluding sewage.

Halophyte a plant that grows in salty soil or with salty water.

Hectare equals 2.47 acres.

Hemicellulose noncellulosic polysaccharides of the cell wall that are easily decomposed by dilute acid, yielding several different simple sugars; principally a polymer of xylose molecules.

Herbaceous plants nonwoody species of vegetation; those lacking a large proportion of lignin, such as grasses.

High-BTU gas a gas containing mostly methane with a heating value of 500–1,000 BTU per standard cubic food (heating value of natural gas = 1,000 BTU/scf).

Hilum a scar on a seed marking the point of attachment of the ovule.

Hydrocarbon a chemical compound containing only hydrogen and carbon.

Hydrolysis the conversion, by reaction with water, of a complex substance into two or more smaller units, such as the conversion of cellulose into smaller sugar units.

Indehiscent seed heads or fruits remaining closed at maturity.

Inflorescence a floral axis with its appendages.

Latex a water emulsion of rubber.

Layering partial covering of a shoot or twig with earth so that it may take root.

Lignin the noncarbohydrate, structural constituent of wood and some other plant tissues, which encrusts the cell walls and cements the cells together.

Lignocellulose plant materials composed of primarily lignin, cellulose, and hemicelluloses.

Lipids water-insoluble biomolecules, such as fats and oils.

Low-BTU gas a gas comprising water, carbon dioxide, carbon monoxide, hydrogen, methane, and nitrogen with a heating value of about 90 to 250 scf.

Medium-BTU gas a gas composed of carbon dioxide, carbon monoxide, hydrogen, methane, and higher hydrocarbons but little or no nitrogen, having a heating value of 250 to 500 Btu/scf.

Mesic plants those usually indigenous to areas with abundant rainfall.

Methane the major component of natural gas. It can be formed by anaerobic digestion of biomass or gasification of coal or biomass.

Metric ton equals 2,204.62 pounds (avoirdupois).

Monoculture the cultivation of a single-species crop.

Municipal solid waste (MSW) the refuse materials collected from urban areas in the form of organic matter, glass, plastics, waste paper, metals, etc., not including human wastes.

Olefin an unsaturated open-chain hydrocarbon containing at least one double bond.

Pappus an appendage or tuft of appendages that crowns the ovary or fruit in various seed plants and functions in dispersal of the fruit.

Parenchyma a tissue of thin-walled storage cells capable of division even when mature and that makes up much of the substance of leaves, stems, and roots.

PVC plastics materials derived from polyvinyl chloride.

Pyrolysis the breaking apart of complex molecules by heating (over the range 390 to 930 degrees F) in the absence of oxygen, producing solid, liquid, and gaseous fuels.

Refuse-derived fuel (RDF) fuel processed from industrial waste, municipal waste, garbage, or sewage sludge.

Renewable energy resources sources of energy that are regenerative or virtually inexhaustible, such as solar, wind, ocean, biomass, municipal wastes, and hydropower energy. Geothermal energy is sometimes also included in the term.

Saccharify to break complex carbohydrates into simple sugars.

Salinize to become salty.

Short rotation intensive culture intensive management and harvesting at two- to ten-year intervals or cycles of specially selected fast-growing species (generally hardwoods) for the purpose of producing wood as an energy feedstock.

Sodic soil clay soil containing sodium chemically bound to clay with few if any other mineral ions. Such clay is too crusty to admit air and water easily and is unsuitable for growing crops.

Stoma one of the minute openings in the epidermis of a plant through which gaseous interchanges takes place.

Syngas a gaseous fuel composed of mainly carbon monoxide and hydrogen produced through heating biomass.

Tar a complex mixture of dark colored material with high BTU value. It may be either a liquid or solid.

Thermochemical conversion the use of heat to change substances chemically to produce energy products.

Trans-esterified oil jojoba oil chemically is an ester. When exposed to a catalyst under appropriate conditions an ester group exchange takes place resulting in a more viscous product.

Xerophyte a plant adapted for life and growth with a limited water supply.

Xylose a five-carbon sugar, a product of hydrolysis of hemicellulose; naturally occurs in many plants and trees, especially hardwood.

Yeast an important example is the single-cell microorganisms that produce alcohol and carbon dioxide under anaerobic conditions; the microorganism that is capable of changing sugar to alcohol by fermentation.

Index